Linking LANs

Titles in the McGraw-Hill Series on Computer Communications

In order to receive additional information on these or any other McGraw-Hill titles, in the United States please call 1-800-822-8158. In other countries, contact your local McGraw-Hill representative.

Linking LANs

Second Edition

Stan Schatt

McGraw-Hill, Inc.

New York San Francisco Washington, D.C. Auckland Bogotá
Caracas Lisbon London Madrid Mexico City Milan
Montreal New Delhi San Juan Singapore
Sydney Tokyo Toronto

Library of Congress Cataloging-in-Publication Data

Schatt, Stan.
   Linking LANs / by Stan Schatt.—2nd ed.
     p.   cm.
   Includes index.
   ISBN 0-07-057063-9
   1. Local area networks (Computer networks) I. Title.
  TK5105.7.S36 1995
  004.6'8—dc20   94-36811
                                       CIP

hc   1 2 3 4 5 6 7 8 9 DOC/DOC 9 9 8 7 6 5

ISBN 0-07-057063-9

The sponsoring editor for this book was Jennifer Holt DiGiovanna, the book editor was Kellie Hagan, and the executive editor was Robert E. Ostrander. The director of production was Katherine G. Brown. This book was set in ITC Century Light. It was composed in Blue Ridge Summit, Pa.

Printed and bound by R.R. Donnelley & Sons.

MH95
0570639

This book, as is the case of all my books, is dedicated to my wife, Jane. In her first-grade class at Park Dale Lane School in Encinitas, California, she teaches children all the really important life lessons, as well as how to read. Every day she makes the world a better place for everyone she touches, especially me.

Contents

Introduction

The first edition of *Linking LANs* was published in 1991. In retrospect, most industry experts have designated this year as the "year of the LAN" because the local area network market took off that year. Novell's NetWare emerged as a de facto standard for network operating systems and PCs finally offered the power and performance necessary to function as file servers. The first edition of this book pointed toward an increasingly complex enterprise network environment, with LAN-to-LAN as well as LAN-to-mainframe and LAN-to-WAN links. (The terms *enterprise network* and *internet* refer to several connected LANs. They're used interchangeably in the book.) The book also recognized the beginning of a convergence of data communications and telecommunications.

This new edition describes today's networking world, where a premium is placed on seamless connectivity. To make the network manager's job even harder, there are several new technologies offering high-bandwidth transmission of information. Should the company go with 100-Mbps Ethernet, switched Ethernet, token ring, or the old tried and true fiber-distributed data interface (FDDI)? What about asynchronous transfer mode (ATM), a transmission technology that promises the unbelievable speed of 600 Mbps?

The first chapter looks at today's complex enterprise network environment. When is a 100-Mbps Ethernet appropriate? Are there interoperability problems with linking high-speed networks to traditional or "legacy" LANs? The chapter points out that enterprise LANs are beginning to resemble old mainframes. In fact, the LAN's glass wiring closet has replacing the "glass house" that was the mainframe's domain.

One reason it's difficult to keep up with network technology is because of the existence of numerous protocols, all of which must live together harmoniously. Chapter 2 looks at several major network protocols, including TCP/IP, and describes them in understandable terms without assuming any previous knowledge of this rather obscure topic.

The third and fourth chapters examine how local bridges, routers, and brouters function. Network managers must often choose between the simplicity and low cost of bridges and the benefits of the more expensive but more versatile routers and brouters. These two chapters compare and contrast these different types of devices and explain when each is appropriate. There are also handy checklists of questions to ask in order to determine which vendor offers the right product for your company.

Chapter 5 examines a very exciting new technology that's likely to become a part of a majority of enterprise environments: wireless networks. Both local area and wide area networks are discussed, as well as specific products that serve as examples of the latest wireless technology.

Enterprise networks often include mainframe computers. Unfortunately, many LAN managers learned about computers via PCs and have very little knowledge of the mainframe world. Chapter 6 describes systems network architecture (SNA), including the various ways that LANs and mainframes can communicate. This chapter serves as very basic introduction to mainframe devices and terminology.

The next three chapters (7, 8, and 9) reflect the convergence of telecommunications and data communications, and illustrate just how important it is to understand both worlds. Chapter 7 describes the world of wide area networks, including public switched networks as well as leased lines. This material will be new to many PC-oriented network managers. Because of all the WAN options available since the breakup of Ma Bell, it's crucial that network managers understand the trade-offs in cost and benefits of using leased T-1 lines, for example, versus switched 56 dial-up or frame relay service to link a corporate office to a branch office. Chapter 8 focuses on integrated digital network (ISDN), the high-speed, dial-up service that has long been touted as the low-cost, high-speed solution for WANs. This chapter answers the question, "Will ISDN finally take off and become the dominant WAN technology?"

Asynchronous transfer mode (ATM) is probably the single most discussed high-bandwidth technology among network managers, yet very few people really understand how it works and what benefits it offers. Chapter 9 takes a close look at ATM and provides detailed information on not only how this technology works but also how it will probably evolve.

The trend toward linking branch offices to LANs at corporate headquarters has resulted in an explosion of low-cost branch-office routers and hubs. Chapter 10 examines the nature of the branch-office LAN and looks at specific products that can link it to corporate headquarters. Once again, there are many different ways to perform this task, and it's important that network managers understand the various options.

The final two chapters in this book look at network management and messaging, two key topics in an enterprise network environment. How should networks running on different platforms be managed from a single console? What are the limitations of simple network management protocol (SNMP)? What will be the result of the specifications developed by the Desktop Management Task Force (DMTF) and how will it affect the enterprise network? These are some of the key questions concerning network-wide directories and e-mail systems.

I wrote my first networking book in 1986. Networks have since changed dramatically, and they continue to change almost monthly. I hope you enjoy this book, and I hope it helps you keep up with this very exciting field.

1

Today's Enterprise Network Environment

In this chapter, you'll explore:

- Mixed topology environments
- Emerging high-bandwidth technologies
- Coexisting server-based and peer-to-peer NOSes
- Enterprise network printers
- Downsizing on enterprise networks

A *local area network* (LAN) is a group of computers that share hardware and software resources at the same physical location. While I assume you're familiar with the concept of a LAN, I don't assume you have more than just a nodding acquaintance with LAN technology. This chapter examines the complexities associated with an enterprise-wide LAN, one that might contain several different networks and use a number of different network architectures, or topologies. Network managers are faced with a daunting challenge—to make everything work together despite the fact that much of the technology was originally designed to be proprietary and noninteroperable.

Network Topologies Found on Today's Networks

Network managers tend to select a network architecture or topology appropriate for a specific function; consequently, many larger companies need to incorporate several different network topologies. Today's enterprise network might include some of the newer high-bandwidth topologies, such as a version of 100-Mbps Ethernet. Take a few moments to familiarize yourself with the major network architectures. After all, you can't tell the players without a scorecard!

Ethernet

May 22, 1973 doesn't quite have the ring of July 4, 1776, but it's a significant day in network history. On that date, Robert Metcalfe wrote a memo at Xerox's Palo Alto Research Center, using the word *ethernet* to explain the principles behind a new type of local area network. That same year Xerox began producing Ethernet network interface cards for its Alto PC.

Intel provided the chips necessary for the network hardware, Xerox provided the software, and Digital Equipment Corporation (DEC) was prepared to run this new network on its minicomputers. In September of 1980, these three companies released a set of specifications for Ethernet that are now referred to as Ethernet version 1. A second version of Ethernet (version 2) was released in November of 1982. Both versions still exist in the field, and there are significant differences between them. The line idle state was changed from 0.7 volts in version 1 to zero volts in version 2. The interface coupling specifications of these two versions also differ. The key point to remember if you ever inherit a legacy LAN with a mixture of network interface cards is that a controller designed for one version won't work with a transceiver designed for the other version.

Traditional Ethernet is associated with the version 1 specifications, which define the network's physical medium (thick coaxial cabling), access control method (carrier sense and multiple access, with collision detection CSMA/CD), and speed (10 megabits per second, or Mbps). The set of specifications also describes the size and contents of an Ethernet packet (72 bytes to 1,526 bytes) as well as the method of encoding data (Manchester).

Notice the maximum size of the packet. Ethernet is a product of its time. It was developed for short bursts of data exchange. It functions very well in an environment where there's constant, heavy network traffic. Think of an amusement park that tries to shuttle large crowds of people around using very small shuttle buses. People would grow more and more impatient waiting in line, and accidents would probably occur.

Shortly after the release of Ethernet, a committee of the Institute of Electrical and Electronics Engineers (IEEE) began deliberation on the development of a set of international, nonproprietary standards for LANs. Given the industry prominence of the three major founders of Ethernet, it should come as no surprise that one standard, IEEE 802.3, is so close to Ethernet version 2 that it's often described as the "Ethernet standard" even though there are differences, described later in this chapter.

Carrier Sense Multiple Access with Collision Detection (CSMA/CD)

The IEEE 802.3 standard contains a protocol (or set of rules) virtually identical to Ethernet's for describing how multiple workstations can access a network when they need to transmit information. Carrier sense multiple access with collision detection (CSMA/CD) dictates that a workstation that wants to use a network must first listen to the network to see if it's busy. If it doesn't detect any signals, it can begin transmitting its message. The workstation then continues to listen for possible network collisions while sending its message.

If a collision is detected, the sending workstation backs off and transmits a special signal to let network users know that a collision has taken place. The receiving station normally discards the contents of the partial message it has received, and all network workstations wait a certain randomly selected amount of time before any station begins transmitting again. Each network interface card is programmed for a different amount of time. This time increases if a collision occurs the very next time the same message is transmitted.

Collisions are inevitable with an Ethernet or 802.3 network because of the nature of CSMA/CD. There's a time interval between when a workstation listens to see if any other workstation is using the network and when it actually begins transmitting. It's entirely possible that a workstation further down the network has begun sending a message but that the signal hasn't yet reached its destination. So workstations in both IEEE 802.3 and Ethernet networks broadcasts their messages in both directions. Figure 1.1 illustrates this method of transmission.

This Ethernet-type standard describes a *contention* network, in which more than one workstation must contend or compete for use of the network. While collisions are inevitable given the network's architecture, the designers assume that the 10-Mbps transmission speed ensures that, even with repeated collisions, users won't experience any noticeable delay. When the number of collisions does cause noticeable delays in response time, it becomes a network management issue, a topic discussed in chapter 11.

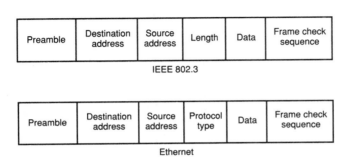

Figure 1.1   Ethernet and IEEE 802.3 frame formats.

Differences between the 802.3 and Ethernet frame format

Figure 1.1 displays the 802.3 frame format. The preamble consists of 56 bits of alternating 1s and 0s, which provides a method for two networks to synchronize. The start frame delimiter (10101011) signifies the beginning of a frame of information, the destination and source fields come from the LLC frame, and the length field indicates the length of the information field that follows it. Data can range from 46 to 1,500 octets in length. The information field is followed by a pad field, which ensures that the frame meets the minimum 802.3 length standards. Finally, a frame check sequence field provides information for error checking.

The major difference between the 802.3 frame and the traditional Ethernet frame is that Ethernet doesn't have a two-byte length field because the length is fixed. Instead, it has a two-byte protocol type field, which indicates which higher-layer protocol, such as transmission control protocol/internet protocol (TCP/IP), is used in the data field. Mixing and matching Ethernet and 802.3 transceivers (devices that actually transmit data from the network interface cards to the media) will result in network errors since an 802.3 or Ethernet node will misinterpret a message intended for the other type of device. Pin layouts are different for Ethernet and 802.3 transceivers. Ignoring this distinction will often result in 802.3 nodes becoming overloaded while handling broadcast Ethernet messages, and users' screens freezing. Guess who the end users are going to ask for help?

802.3 as an evolving standard

While Ethernet specifies only 50-ohm coaxial cabling, the IEEE 802.3 standard currently supports several different types of transmission, including baseband and broadband coaxial cable, and twisted-pair wire. The cabling will vary according to the recommended maximum distance. A now defunct product known as StarLAN was at one time offered by a number of vendors, including AT&T. This particular version of Ethernet provided a 1-Mbps transmission speed over 500 meters (1base5). Thick coaxial cable (50-ohm) has a 500-meter limit, so the 802.3 standard is referred to as 10base5, meaning 10-Mbps baseband coaxial cabling with a 500-meter limit. Thin coaxial cable can carry a signal 185 meters (10base2, or "cheapernet"), while unshielded twisted-pair cable (10baseT) is recommended up to 100 meters.

The older StarLAN 802.3 specification for a 1-Mbps network with a 500-meter maximum length is known as 1Base5. (I'll explain 10BaseT shortly.) Because the IEEE 802 subcommittees continue to meet as new technology develops, the standards continue to evolve. The 802 standards use a layered set of protocols very similar to the OSI model, so it's possible to add to the MAC layer without having to change anything in the LLC layer. Ethernet's flexibility will become apparent when you look at a 100-Mbps version of this topology a bit later in the chapter.

A closer look at an Ethernet network

Before looking at some variations, let's examine the components of the traditional Ethernet or 802.3 LAN; just as the Colt 45 became known as "the gun that won the West," Ethernet is "the LAN that made local area networks respectable." Systems managers could install an Ethernet network and depend on it without worrying about being fired for buying fringe technology.

An Ethernet workstation contains an Ethernet-specific network interface card (NIC). This NIC is responsible for handling collision management and encoding and decoding signals. With thin coaxial cabling, a transceiver is part of this NIC (as illustrated in Figure 1.2), while terminators are external on thick coaxial networks. The terminators complete the segment's electrical circuit.

A single segment can hold a maximum of 100 workstations, and multiple segments can be linked by repeaters. Up to 1,024 workstations can exist on a single Ethernet network, with a maximum of two repeaters in the path between any two stations.

The transceiver actually generates the electrical signals to the coaxial cabling and maintains their quality. Conversely, transceivers are also responsible for receiving network signals and detecting packet collisions. When a transceiver on the transmission side detects a collision, its NIC link management function turns on a collision detect signal. The transmission side of link management in turn sends a bit sequence referred to as a *jam*. This jam signal causes transmitting workstations to terminate their transmission, and their NICs randomly schedule the retransmission. Meanwhile, the NICs of the receiving workstations have been examining damaged or fragmentary packets and discarding the "runts."

Ethernet transceivers internally generate a signal quality error (SQE), which is often referred to as a "heartbeat." This is read by the NIC to ensure

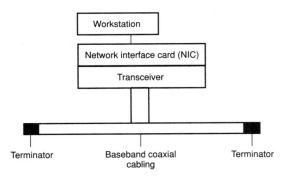

Figure 1.2 Ethernet architecture.

that the transceiver is working properly, sort of like having your spouse hum every couple of minutes so you know he or she is alive and listening to you. Unfortunately, the heartbeats for 802.3 and Ethernet transceivers use different timing. Some manufacturers provide transceivers that can be manually set for one standard or the other.

Because Ethernet broadcasts in all directions, there are cable-length limitations, as discussed earlier in this chapter. Original Ethernet, with its thick coaxial cabling, permitted one 500-meter segment with a maximum of three segments joined together by repeaters to rebroadcast the packets on the network.

Other limitations for 10Base5 include a maximum of 100 transceivers for a single cable segment, with transceivers spaced at least 2.5 meters apart. The 10Base5 specification describes drop cables connecting the workstation's NIC to transceivers on the bus. A break in a drop cable wouldn't bring down the entire network, but only the affected workstation. On new "cheapernet" networks, however, the transceivers are built into the NIC. A break in cabling will bring down the entire network.

Besides cabling breaks, other common equipment-related Ethernet problems include "jabber" and individual station failures. A malfunctioning transceiver can "jabber" (send out continuous streams of packets). This problem can be identified with a protocol analyzer. Similarly, a malfunctioning workstation that sends packets below the minimum accepted length ("runts"), or packets that are continually of the maximum size with no variation (causing traffic congestion) can also be spotted with a protocol analyzer.

A bus by any other name . . .

As you saw earlier in this chapter, Ethernet and 802.3 are bus networks. A *bus* is a data highway that's laid in straight sections known as *segments*. Segments can be linked together, and repeaters can be used to extend a network's size.

The StarLAN and 10BaseT topologies are distributed stars. Physically, as shown in Figure 1.3, such networks have workstation cabling that radiates from concentrators in a star-like arrangement. Logically, though, they still operate as bus networks.

Sometimes known as *passive stars*, these networks enjoy the benefits of a star-like architecture since it's possible to maintain better network management and control. Bypass circuitry means that if one workstation fails, it can be bypassed by other workstations in the star by the concentrator or multiport transceiver from which the workstations' cables radiate.

10BaseT

SynOptics Communications was the first company to release an Ethernet product running at 10 Mbps over unshielded twisted-pair cables in 1987. It

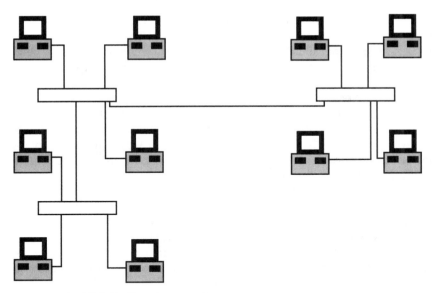

Figure 1.3   An 802.3 distributed star topology.

was followed by products from several other companies, including David Systems, Hewlett-Packard, 3Com, and DEC. Unfortunately, each company had its own proprietary set of specifications that were incompatible with those of the other companies. Large companies were reluctant to invest heavily in a small company's proprietary technology, since it might very well lead to a dead end should the small company fail. The 802.3 committee ameliorated these fears with its 10BaseT standard.

10BaseT has a 100-meter maximum segment length, a link-integrity test specification, and the ability to disconnect a segment if there's a failure without bringing down the entire network. A 10BaseT description of the media access unit (MAU) indicates several new features not found in other 802.3 specifications. It has a jabber function that disables the MAU if it transmits after an maximum allowable time period. After being "gagged" for a period of time, the MAU will enable itself and try talking to the network again. A workstation signal quality error (SQE) test monitors the health of each MAU and ensures that workstations are able to participate as receivers and senders of network packets. Repeaters under 10BaseT are able to disconnect a malfunctioning MAU without disconnecting all other workstations. When the failure is corrected, the specified port that has been taken out of service is reconnected to the network.

Another useful 10BaseT feature is "intelligent squelch," which means that 10BaseT can function in an environment containing a wide range of conflicting signals, including voice traffic, the new integrated services digital network (ISDN) traffic, and asynchronous data traffic. Like a mother who's able to pick out her baby's cry even in a crowded room full of crying babies, intelligent squelch filters out other signals so Ethernet signals can be detected.

Another problem that 10BaseT was forced to address is the distortion of a signal as it travels over a twisted-pair wire. This problem long prevented high-speed Ethernet traffic over this medium. A pre-equalization technique was developed in which the signal is distorted before it travels to compensate for the distortion that takes place during transmission. The signal reaches its destination in very much the same form as it began.

The major advantages of 10BaseT is that so much unshielded twisted-pair wire is already installed, and that installation is easier and connections are more reliable than with the coaxial cable version of Ethernet. While new Ethernet installations in the U.S. are over 90 percent 10baseT, Europe has been much slower to adopt this technology. The major reason I've spent so much time discussing coaxial cable versions of Ethernet is that network managers are likely to inherit complex legacy LANs containing a wide range of cabling and topologies. The only concern I have with 10BaseT technology is that network managers might try to save money by simply assuming that existing cabling within an older building will be adequate. Buildings that appear to have a lot of extra unshielded twisted-pair cables might have poor assignment records so it's almost impossible to trace back all pairs. There might also be bridge taps from pairs tied together somewhere inside the building. Some loops might hang in the air—roads that lead nowhere. Conduits might be blocked, and pairs broken or poorly spliced together.

Ethernet on fiber-optic cables

It's possible to use fiber-optic cabling on an 802.3 network. The major advantages of this approach are immunity to any kind of electrical interference and the distance you can cover. The fiber-optic links can be up to 4.5 Km in length. Codenoll, one of the leading vendors in this market, points to its installation of the world's largest fiber-optic network, a 1.5-million-square-foot, 44-story network with over 3,000 stations and 92 miles of fiber at Southwestern Bell's headquarters in St. Louis, Missouri.

Each network workstation must have a NIC designed for 802.3 transmission over a fiber-optic network. Codenoll offers an external transceiver alternative. In any event, the transmitters in these products convert electrical signals to pulses of light while the receivers convert the light-wave signals back into electrical signals.

An optical bus star coupler sends optical signals to each station on the network. They're the equivalent of the concentrator or hubs you saw on the 10BaseT standard. Repeaters extend distances; they also permit "cascading stars" by connecting together different optical star couplers. Using Codenoll's products as an example, these couplers are available in coaxial-to-fiber, fiber-to-fiber, and coaxial-to-coaxial models. The actual fiber cables come with connectors preattached, and replace coaxial cable and twisted-pair wiring. Figure 1.4 illustrates a fiber-optic Ethernet network.

Faster versions of Ethernet

Many companies with large Ethernet LANs have begun to reach the network equivalent of traffic gridlock. Once the utilization starts to climb over 40 percent, throughput drops and users begin to complain. Network managers have begun to look at several ways to speed-up traffic without having to add an entirely new network freeway.

Switched Ethernet. Kalpana was the first company to introduce the concept of switched Ethernet. Other companies, such as Alantec and Artel, have followed. This technology divides a large network into smaller segments with fewer users on each segment. Each switch port is responsible for filtering traffic sent to its attached segment. If a node on one segment transmits a message to a node on another segment, then the port forwards the message to the switch fabric and the appropriate destination port. The switch supports simultaneous 10-Mbps connections between segments.

Kalpana uses a technique known as *cut-through* rather than buffered switching to transmit packets. A port in the switch transmits a packet to its destination port immediately upon reading its destination address. This approach results in the least amount of latency in transmission between ports.

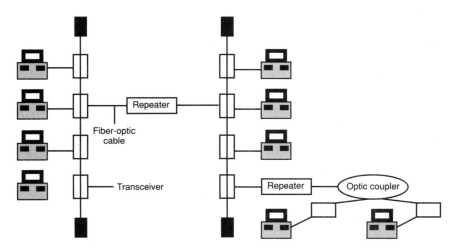

Figure 1.4 A fiber-optic Ethernet network.

The negative consequences of this approach is that there can be packet collisions as well as bad packets arriving at the destination segment. On most other switches, the switch fabric usually consists of some form of shared memory. A packet is received into this memory and its final destination port is determined from an address table using a microprocessor and some built-in software.

An Ethernet switch is advantageous as a short-term solution, but switches have a finite number of ports so purchasing a switch presents the same problems on a smaller scale as purchasing a PBX telephone system. A network manager must look at the present network's projected growth and try to determine if there's a switch available that provides the right level of capacity. A company with 22 segments and the prospect of adding two to three additional segments isn't a good candidate for a switch if the two options available are 24 or 48 nodes.

Full-duplex Ethernet.   In late 1993, Kalpana introduced full-duplex Ethernet technology. It consisted of two 10-Mbps channels, one for receiving data and one for sending data in a point-to-point connection. Both ends of a full-duplex connection can simultaneously send and receive data via a null modem cable, resulting in a maximum aggregate of 20 Mbps. This technology has grown beyond the limitation of a Kalpana switch and now includes bidirectional Ethernet adapters. A server using a Compaq NetFlex-2 EISA adapter or IBM's EtherStreamer-32 micro channel adapter can communicate at 20 Mbps with a Kalpana switch.

There's a major performance limitation to full-duplex Ethernet. The only way this technology can achieve close to 20-Mbps speed is when traffic is balanced in both directions. Since most client/server communication is primarily one-way, it's likely that overall performance will fall below expectations. Cards using this technology do provide far higher throughput even at half-duplex mode, however, so network managers should still consider full-duplex Ethernet as a tool for improving overall network efficiency.

Full-duplex Ethernet is a switched, dedicated version of standard Ethernet, in which 10-Mbps channels can be established bidirectionally for an aggregate throughput of 20 Mbps. Because its Micro Channel bus contains a burst mode, IBM is positioning its LANStreamer and EtherStreamer boards as the best products for customers interested in this technology. Texas Instruments is interested in full-duplex Ethernet as a way to differentiate its products from those of other Ethernet vendors. SynOptics has offered this type of product via its collaboration with Kalpana on full-duplex switched integrated within its hubs. Compaq is also a player in this market, with a NetFlex board using Texas Instruments' chips.

The problem for network managers is the lack of interoperability among these different approaches. While Cabletron has a been a leader in gathering support for interoperability testing, many vendors are taking a wait-and-see attitude; they aren't sure how much interest customers will have in a 20-Mbps topology. Unless a customer is already locked into a proprietary intelligent product, such as those offered by Cabletron and SynOptics, it's probably wise to think twice before committing resources to this technology since it doesn't offer the interoperability that most network managers seek for enterprise-wide networks. Also, at a cost of approximately $700 per port, full-duplex Ethernet is priced far above 100-Mbps Ethernet.

100-VG AnyLAN. Hewlett Packard, AT&T, and IBM have been the driving force behind 100BaseVG AnyLAN, a 100-Mbps version of Ethernet/token ring running on UTP from cable grade 3 to 5. Ultimately, the topology will become an IEEE 802.12 standard. The 100-VG (voice grade) specification includes support for shielded twisted-pair wire (STP) as well as fiber. Because so many token-ring customers have large STP wiring plants, they're candidates for this new technology.

The 100-VG technology uses a new signalling system known as *demand priority* rather than Ethernet's traditional CSMA/CD. Demand priority is deterministic, so equal access is provided to each node. The hub scans each port to test for a transmission request, and then grants the request based on priority. There are two different levels of priority available, high and low.

Demand priority operates over four-pair, category 3, 4, or 5 unshielded twisted-pair wire (UTP), two-pair shielded twisted-pair (STP or IBM type 1) cable, and single-mode or multimode fiber cabling. Transmission over unshielded twisted-pair wire uses a technology known as *quartet coding*, illustrated in Figure 1.5. Data is split into four parallel streams, with each stream directed down each pair of a four-pair UTP cable. On each pair of

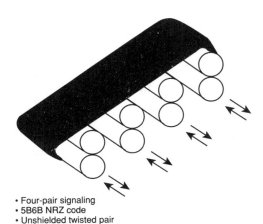

• Four-pair signaling
• 5B6B NRZ code
• Unshielded twisted pair

Figure 1.5 Quartet coding.

wires an efficient 5B6B NRZ encoding scheme is used to transmit two bits of information per cycle. In this way, quartet coding permits 100 Mbps worth of data to be sent across a four-pair UTP cable while keeping individual signal frequencies at no more than 15 MHz, well below U.S. FCC limits.

To transmit 100 Mbps worth of data over shielded twisted-pair cable (a two-pair STP media such as IBM type 1), 100-VG AnyLAN transmits the data in two parallel streams. This approach takes advantage of the comparably higher level of shielding provided by STP to transmit at higher frequencies, achieving 100-Mbps transmission speed using only two pairs of wire.

Like 10BaseT, multiple 100BaseVG AnyLAN hubs can be cascaded within a single subnet to extend the topology without requiring additional bridges or other components. In a cascaded 100-VG AnyLAN configuration, the demand priority protocol allows hubs to automatically recognize whether or not they're connected to a higher-level hub. When a lower-level hub receives a packet request from a connected node, it forwards the request to the next-higher-level hub before acknowledging the requesting node. The top-level hub arbitrates this request along with other packet requests from other nodes or hubs. When the top-level hub acknowledges each request in turn, the acknowledgment cascades down to the lower-level hub, which then acknowledges each of its own pending requests before relinquishing control back to the higher-level hub. As the lower-level hub passes the acknowledgment to the request node, the node then transmits its packet with an assurance of uncontested transmission throughout all the connected hubs in the subnet.

In this way, the demand priority arbitration scheme is extended across multiple hubs with no loss of fairness or efficiency. As in 10BaseT, a 100-VG AnyLAN network can also be segmented with bridges or switches to allow simultaneous packet transmissions on separate subnets, further increasing the bandwidth available to individual nodes or servers. Figure 1.6 illustrates 100-VG AnyLAN topology.

One major criticism of 100-VG technology has been its break with traditional Ethernet CSMA/CD network access and a perceived lack of compatibility with existing Ethernet installations. When 100-VG AnyLAN technology is used to upgrade portions of an existing 10BaseT Ethernet network, a speed-matching bridge is needed to connect the 10BaseT and 100-VG AnyLAN subnets. The bridge buffers the higher-speed packets as they enter the slower-speed network. Since the same Ethernet packet format can be used on both the 10BaseT and 100BaseVG AnyLAN subnets, no packet translation or other processing is required.

To upgrade 10BaseT nodes, you must replace their LAN adapters with a 100-VG AnyLAN adapter. No new cabling needs to be installed. The same RJ-45 connector and unshielded twisted-pair cabling used for 10BaseT LANs are acceptable. The second step in replacing some existing 10BaseT with 100-VG AnyLAN nodes consists of disconnecting the nodes' cabling

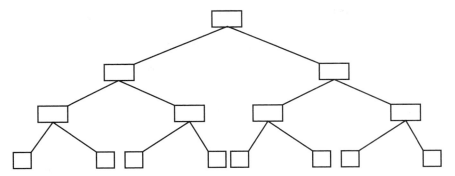

Figure 1.6 100VG AnyLAN topology.

connectors from their 10BaseT hub ports within a wiring closet and recon-
necting them to 100-VG AnyLAN hub ports.

Fast Ethernet or 100BaseX. Grand Junction shocked the networking community in September 1992 when it announced a scheme to achieve 100-
Mbps transmission speed with Ethernet while retaining the topology's
CSMA/CD approach to network access. Such a product would mean that
current software drivers for Ethernet would work without change. Late
1993 saw the first delivery of products, while the IEEE 802.3 Committee
still debated specifications for 10BaseX. The IEEE has turned over development of final specifications to its 802.30 Committee.

The Grand Junction proposal for fast Ethernet is predicated on a
CSMA/CD media access control (MAC) layer used in conjunction with the
ANSI X3T9.5 physical medium dependent (PMD) layer. Two pair of UTP
data-grade cabling is required. The result is that the technology speeds up
the frequency of packets transmitted without changing the structure of the
packets themselves.

The major advantage of 100BaseX over any alternative approaches to 100-
Mbps Ethernet is that the technology is compatible enough with existing
Ethernet to integrate into existing Ethernet LANs via bridging or dual-speed
network adapters. There's a considerable problem, though, for network
managers who don't have category 5 wiring already installed. It's likely that
companies will purchase intelligent hubs that offer both 10-Mbps and 100-
Mbps Ethernet. Rather than purchase 100-Mbps adapter cards for all nodes,
network managers are likely to initially link together servers at 100 Mbps.
Table 1.1 summarizes features for the two types of 100-Mbps Ethernet.

Which 100-Mbps Ethernet topology should you choose?

There are several factors for network managers to consider when selecting
between 100BaseVG and 100BaseX. If a company is planning to upgrade

TABLE 1.1 Two Types of 100-Mbps Ethernet

Features	100Base-VG	100Base-X
Signaling	Quartet signaling using all four pairs of wires to send or receive data	Duet signaling using Ethernet's traditional CSMA/CD scheme, requiring one pair of wires to send data and another pair to receive data
Distance	100 meters	100 meters
Traffic prioritization	Demand priority assigns high or low priority to each packet	None
Wiring	Category 3 voice-grade, unshielded, twisted-pair wire, fiber, and IBM Type 1 shielded, twisted-pair wire	Category 5 data-grade UTP

only selected nodes to 100-Mbps Ethernet, then the type of existing cabling isn't as significant an issue as if all nodes need to be upgraded. Replacing all unshielded twisted-pair wire to data-grade wire in order to run 100BaseX could prove expensive for all nodes. A second major consideration is whether real-time video will be transmitted over the network. The low latency and guaranteed delivery found with 100-VG make it a better topology for this type of video. A third consideration is how heavy the traffic will be on the 100-Mbps Ethernet LAN. The 100BaseX topology is still a CSMA/CD contention network; heavy traffic will still result in data collisions and less network throughput.

Finally, a major consideration in selecting a 100-Mbps topology is how much current enterprise network interoperability focuses on the traditional Ethernet packet format. How available are 100-VG Ethernet network drivers for all the software currently running under traditional Ethernet?

IEEE 802.5: Token-Ring Networks

Many large corporations have a mixture of LAN topologies. While Ethernet does dominate, there are a good deal of token-ring networks found in these establishments. In fact, over half of all Fortune 1000 companies have both token-ring and Ethernet LANs coexisting at the same sites. The IEEE 802 committee developed a standard that became known as IEEE 802.5 for a noncontention local area network, a network that's a logical ring but a physical star. Cables to individual workstations radiate from wire concentrators known as *multistation access units*. The IEEE 802.5 committee published its set of specifications in the "blue book" and has been adding to this standard on a regular basis. The original specification described a 4-Mbps transmission speed. Virtually the only adapters sold today are dual-speed 4-Mbps and 16-Mbps adapters. What is surprising is that a significant number of token-ring customers have switchable cards but haven't felt the need to upgrade their multi-station access units as required to migrate to 16-Mbps performance.

Using tokens on 802.5 networks

A *token* consists of a predetermined bit pattern and can be used by only one workstation or network node at a time. The transmitting workstation physically alters this token's bit pattern, which announces to all other work-stations that the token is in use. This method resembles the way a taxi dri-ver alerts potential customers with an "in use" flag on top of the cab. The token, now transformed into a frame of information containing the message to be sent, is transmitted around the ring until it reaches its destination workstation.

Messages sent along a token-ring network are received by each network workstation, which in turn checks to see if it's the frame's correct destina-tion. If not, the workstation acts as a network repeater, retransmitting the frame to the next network workstation. Finally, the destination workstation receives the message and copies it to its internal memory before retrans-mitting the frame back to the sending workstation. The frame makes its way around the ring back to the sending workstation, which observes that the message was successfully copied and then resets the token so it's available for another station to use.

This passing of a token might remind you of a children's game in which children sitting around a circle transmit a message by whispering it, until the child who started the cycle hears the information. What would happen if one of the children fell asleep and couldn't transmit the message? Obviously, the message would fail to come full circle. Later in this chapter you'll see how a multistation access unit or wire concentrator alleviates this problem by bypassing inactive workstations so the continuity of the ring isn't broken.

Basic components of a token-ring network

In this section, you'll look at the key components of a token-ring network, including its network interface card, multistation access unit, and cabling. It's important to recognize that IBM's Token-Ring Network (IBM loves to use capitals as well as abbreviations) differs in some ways from other 802.5 networks. That doesn't necessarily mean IBM's approach is better, but it does mean that you have to be careful not to mix and match NICs with dif-ferent chip sets or different 802.5 cabling schemes.

Network interface cards (NICs)

Workstations must have token-ring network interface cards (NICs). These cards contain the medium access layer protocols that are at the heart of the

802.5 standard. Not all 802.5 NICs are the same; while all 802.5 standard NICs support IBM's source-routing scheme, IBM's NICs actually handle source routing with their firmware and don't require additional software to perform the task.

IBM and Texas Instruments jointly developed an 802.5 chip set to ensure that third-party manufacturers of LAN equipment would provide compatible products. Ironically, virtually all 802.5 NIC manufacturers use this chip set except Ungermann Bass and IBM. Even though dozens of vendors offer 802.5 NICs, there are significant differences. One major difference is in the way the cards handle memory. The three major approaches toward handling memory are shared memory, direct memory access (DMA), and bus mastering.

With *shared memory*, a portion of the host's memory is mapped to the NIC's memory. The host can read information directly from its own memory, thus providing very fast access. The NIC has jumper or switch settings to assign the buffers on the card to a system memory address, and the CPU reads the RAM information on the card just as if the information was contained in high-memory RAM on the motherboard. There is a drawback, however. How much host memory did the manufacturer allocate? Too little memory allocation will result in the need for additional communications and transmissions, which will slow down operations. Also, some information could be lost during this process. Too much memory allocation could cause conflicts with other key operations. The normal address range used by the chip set for this function could create conflicts with other functions that use this range of memory, including system video memory and the extended memory specification (EMS) paging area.

Direct memory access (DMA) offers an alternative memory approach. A DMA controller residing on the NIC's motherboard assumes responsibility for determining the source and destination addresses of the data to be moved along the data bus. It requests and receives permission to use the data bus and then performs the necessary read and write operations. Unfortunately, on PCs the DMA speed remains what it was on the original Intel 8088 machines, 4.77 MHz. This approach isn't as fast or as efficient as shared memory. It's reasonably efficient when used for large data transfers, but very inefficient for small data transfers because of the overhead. As a systems integrator, your ability to determine what kinds of data transfers take place routinely will help you decide whether the DMA approach is acceptable for the network you configure.

The third memory approach, *bus mastering*, is the most efficient way to use memory, but you'll need a PS/2 workstation based on the micro channel architecture (MCA) or extended industry standard architecture (EISA) 16-bit bus clones and appropriate NICs to use it. The NIC in effect is the "master of the bus" because it controls the bus and not a DMA controller sitting on a motherboard. The NIC writes information directly to the host's memory without the need for request permission. Also, the NIC needs to perform only a write operation and not the read operation required with the DMA approach. If the network you're configuring has a lot of small data transfers, this approach might be significantly faster for you than DMA.

Just as NICs use different memory schemes, they also offer different-sized buffers. It's particularly important that you select a card with a very large buffer for your file server so it can handle its heavy I/O load. Other significant features that differentiate 802.5 NICs include the included software and the quality of their driver software.

Let's take the included software on a NIC a step further. TI's 802.5 chip set includes basic driver software, but some companies have rewritten these drivers for their NICs using assembly-language routines that speed up the throughput between the operating system and the hardware.

Cabling and multistation access units

The cabling connecting the network can range from type 3 unshielded telephone wire to fiber-optic and IBM data-grade cabling. You won't be able to reach the promised 4-Mbps transmission speed using unshielded telephone wire if there's any interference. Another limitation of using inexpensive telephone wire is that it will support a maximum of only 72 workstations compared to the 260 workstations that can be supported on one ring with data-grade coaxial cabling.

While a token-ring network is a logical ring, it's also a physical star with workstations radiating from multistation access units (MAUs). These MAUs contain bypass circuitry so if a workstation isn't plugged in or is logged off the network, a relay bypasses the workstation's MAU port so that continuity of the ring is not broken. Figure 1.7 illustrates how these MAUs function.

MAUs are usually placed in a centralized wiring closet, and the cable connecting the MAU to a workstation is known as a *lobe*. The maximum distance of these lobes varies according to the type of cabling used with the 4-Mbps version of a token-ring network, which supports a maximum length of 330 feet for twisted-pair telephone wire and 984 feet for data-grade media.

Figure 1.8 illustrates a group of MAUs in a centralized wiring closet. Connections at the wiring closet are made at punchdown blocks or through patch panels. Connections at the workstations are made using standard telephone wall jacks.

MAUs and repeaters

For larger installations, it might be impossible or simply undesirable to place all MAUs in one central wiring closet. Copper repeaters can extend the distance between MAUs by up to approximately 2,500 feet. Figure 1.9 illustrates how these repeaters are linked to the MAUs.

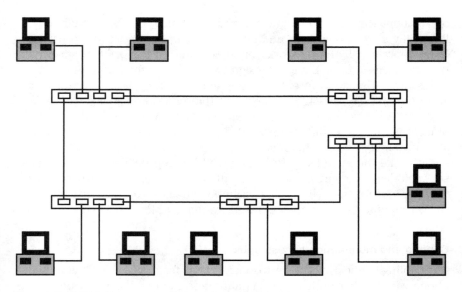

Figure 1.7 Multistation access units in a token-ring network.

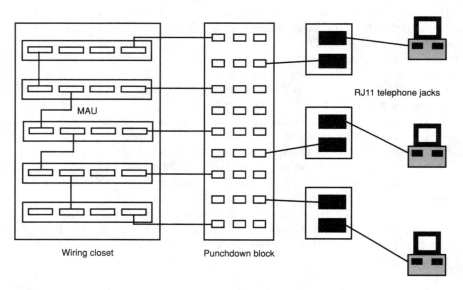

Figure 1.8 MAUs in a wiring closet connected to workstations using a punch-down block.

There are some "rules of thumb" associated with token-ring networks. Generally, you want to minimize the number of wiring closets. With twisted-pair telephone wire, you can have a maximum of 72 workstations using up to nine MAUs located in a single wiring closet as long as all lobe lengths are 1,000 feet or less. If you go beyond this length, you'll need to use repeaters or bridges.

With shielded data cabling (type 1 or 2), a token-ring network can consist of a maximum of 260 workstations with up to 33 MAUs located in a single wiring closet, with all lobe lengths 2,000 feet or less.

The 802.5 frame

The 802.5 frame is quite different from the 802.3 frame described earlier in the chapter. This difference explains why you simply can't plug the two networks together even if they both use the same baseband transmission. Figure 1.10 illustrates the 802.5 frame.

The starting delimiter signifies the beginning of a frame; its unique pattern (JK0JK000) prevents it from being mistaken for data. J and K here represent nondata symbols. The access control field is where you find the priority and reservation bits as well as the monitor bit. A workstation might use any one of eight different priority levels to indicate that it needs to reserve use of a future token. Other workstations will compare their priority levels with this number and defer if they see a higher priority in the field. The monitor field refers to token management, a topic I'll discuss shortly.

The frame control field is used to indicate whether the frame contains logical link control (LLC) data or a medium access control (MAC) control parameters. The destination address and source address fields are self-ex-

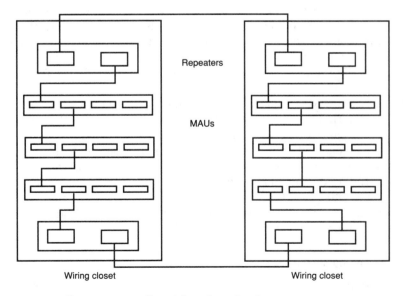

Figure 1.9 Repeaters extending a token-ring network.

Starting delimiter	Control	Destination address	Source address	Information	FCS	EFS	Ending delimiter	Frame status

Figure 1.10 The 802.5 frame.

planatory. The source and destination address fields are designed to convey a number of different kinds of information. If the first bit is set to 1, then the message is a group broadcast for everyone on the network. An initial zero bit, on the other hand, indicates that this is a message addressed to a specific workstation. The second bit in the address field indicates whether the address is global (0) or local (1). Local refers to another node on the same network.

Each workstation has a unique 48-bit address obtained by the manufacturer from IEEE, which is burned into the PC adapter ROM chip. Locally administered addresses are assigned by the local network administrator; these addresses override the universally administered addresses found on the ROM chip. The address fields have been designed to accommodate the addresses of workstations that exist on other rings; in fact, the first two bytes of these two fields are designated for a workstation's ring number. The addressing can become quite complex since the 802.5 committee has proposed an address structure that includes space to indicate multiple rings, bridges, etc., the same way a letter to a college student might require several address lines to indicate the specific college, street address, dormitory, post office box, etc.

The frame check sequence field is used for error checking, while the end of frame sequence and ending delimiter are fields specify the end of a frame. The frame status field indicates whether the frame is "good" or "with error." If the frame is with error, it means one of several things; either the FCS doesn't match, the frame is too large for the buffer space, or the frame is too small to be valid (a runt).

The frame status field is used by the workstation originating the information frame to determine if the workstation designated to receive the message actually recognizes its own address. The originating workstation sets the addressed recognized bit to zero, while any other station on the ring sets it to 1 if it recognizes the address as its own. The originating workstation also sets the frame copied bit to zero. When the receiving workstation copies the frame into a read buffer, it sets the bit to a 1. When the frame returns to the originating workstation, it checks the frame status field to see if the frame was recognized and copied correctly.

The token on an 802.5 network

A token is nothing more than a specific bit pattern that workstations can recognize. Figure 1.11 illustrates the token on an 802.5 network. A workstation that needs to use the network will grab the token when it arrives, and change one bit to transform it from a token into what is known as a start-of-frame sequence (SFS).

Workstations that need to use the token on an 802.5 network can indicate any one of eight different levels of priority while placing a reservation for it. A network workstation releases its token after each transmission because it isn't permitted to broadcast continuously regardless of its high priority. The monitor bit is broadcast as a zero in all frames and tokens, except the monitor itself, in order to ensure that no frame monopolizes the ring and no token has a priority greater than zero. Every workstation thus has equal access to the network.

When a workstation is ready to continue a message's path through the network by rebroadcasting its bit pattern, it examines the reservation bits (RRR). If it has a message of its own that it wants to send and its priority is higher than the present sender, it will raise the value of this three-bit field to its own level, assuring that its message is the highest one waiting to be transmitted. Further down the network, a second workstation with even a higher priority might change this bit pattern to reflect its own priority level and thus "bump" the previous workstation from its reservation for the next available token.

The priority established in the PPP priority field reflects the workstation's priority to use the network. If a workstation observes that its priority is higher than the one already reflected in the RRR reservation field, it will raise the value to its own level and thus reserve the next token for its own use. The address of the bumped workstation will then go into a memory location that serves as a queuing area for displaced workstations seeking access to the network but forced to wait their turns.

The role of the active monitor

Token-ring networks use an active monitor to ensure smooth network operation and handle error conditions. The first workstation to join a ring becomes the monitor, sort of like being president of a country consisting of one citizen. As the monitor, this workstation periodically generates an active monitor present frame, which tells other workstations that there's a monitor alive and functioning. Other workstations periodically issue standby monitor frames, which indicate that they're ready to assume this role if the monitor fails.

Starting delimiter	Access control PPPTMRRR	Ending delimiter

Figure 1.11 An 802.5 token's format.

P = Priority mode   M = Monitor count
T = Token bit   R = Priority reservation

What happens if the monitor *does* fail? Each workstation keeps track of time. When one realizes that it's long past time for the active monitor present frame, it generates a claim token frame. Other workstations begin to realize the same thing, and they too begin to generate claim token frames. Which workstation is going to win the job of active monitor?

Now something very interesting happens, sort of how dogs determine which one is going to rule the neighborhood. Each workstation receives claim token frames issued by other workstations, and it compares its source address with the source address of these claim token frames. If it has a lower source address, the workstation drops out of this contest and begins issuing standby monitor frames. If it has a higher source address, then it ignores this contender's frames and continues generating its own claim token frames.

The frame check sequence field is responsible for error checking the frame check, destination address, source address, and information fields. Bit 7 of this field is the error-detected bit. The workstation originating a frame sets this bit to zero, while the first station to detect a transmission error sets it to 1. The first station to detect this error flag counts the error and prevents other stations from also logging this error. This method helps to localize where an error has taken place.

The workstation serving as the monitor scans the network for transient and permanent errors. Transient errors are logged by various workstations on the network as soft error conditions. These errors can generally be corrected by retransmitting the frame. Permanent errors, on the other hand, can disrupt network operations. If a frame comes back with the indication that a destination workstation hasn't recognized its own address and copied the frame, the monitor workstation can use the bypass circuitry built into the multistation access units to bypass the defective station and maintain the network's integrity.

Error checking

The active monitor examines the frames circulating on the network and removes any of them that are defective, issues a new token, and ensures that the network runs smoothly. It also helps identify and remove defective network nodes. Figure 1.12 illustrates a token-ring network with four workstations. Workstation 1 starts to receive fault messages indicating that a problem exists somewhere between it and its nearest active upstream neighbor (NAUN). Workstation 1 begins issuing media access control (MAC) layer beacon frames that contain its own address as the source address and its NAUN (workstation 4) as the frame's destination address. When workstation 4 receives a total of eight beacon frames, it removes itself from the ring and begins self-testing. It's able to test its own NIC hardware, lobe cabling, and interface to the MAU. If workstation 4 doesn't find any errors, it will

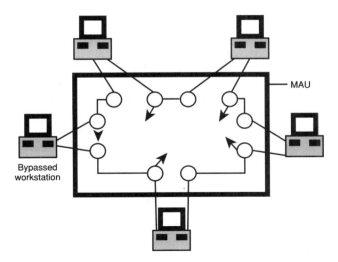

Figure 1.12   A token-ring network with a defective node.

reenter the ring. Note that when a workstation is powered on, the NIC initiates a self-test that checks its memory and circuitry. If an error is found, the workstation won't enter the ring until it's corrected.

Meanwhile, workstation 1 is still receiving error messages. Does it know that workstation 4 isn't the guilty party? Horror of horrors, could it be the cause of these error messages? Workstation 1 takes itself out of service and begins self-testing. If it finds nothing wrong, it will return to the ring. Since workstation 1 was the active monitor, the remaining workstations will contend for the job using the method described earlier.

The functions of the monitor workstation are provided by the IBM Token-Ring Network manager program. It provides continuous monitoring of the network for errors as well as the corrective actions I've been discussing. It also logs network error and status information for report generation. This program provides the mechanism to create addresses for each workstation, as well as establish passwords.

Common problems on token-ring networks

When examining Ethernet, you saw that there were a number of common network problems, including cable breaks and "jabber" conditions. Many systems integrators swear that, even with all of its self-diagnostics, token rings are far more difficult to maintain than Ethernets. 802.5 networks also have their cable breaks, and it's possible for the relays that take a workstation out of service on an MAU to stick so the integrity of the ring is broken. Sometimes NICs will garble or abbreviate frames. There are many reasons why this can happen, including electrical interference, low batteries, NICs going bad, and loose connectors.

Sometimes on token-ring networks a token is accidentally duplicated or lost. Two workstations might believe they hold the token and then attempt to transmit a frame simultaneously. This situation is handled very nicely by the network. Workstations examine the source address field of a returning frame and make sure the address matches their own address. If the two addresses don't match, the station aborts its own transmission and doesn't issue a new token.

16-Mbps token ring

Since companies who install 16-Mbps token rings are usually the companies with very large networks, fiber optics play an important role in such networks' design. Many vendors offer conversion from optical fiber to token ring. These converters permit optical links between wiring closets with 62.5/125 or 100/140 multimode fiber-optic cabling.

These fiber-optic systems often provide network management software as well as two distinct signal paths. One path is in the primary direction of the network, while the other path is in the backup direction. Normally, the primary path carries the network traffic and the backup path carries ring maintenance signals. The optical converters are intelligent enough to remove themselves from the ring if they sense that they're misfunctioning, and then later reinsert themselves if they no longer identify a problem.

Early token release on the 16-Mbps token ring

A workstation on a 16-Mbps token-ring network is permitted to transmit a token immediately after sending a frame of data instead of waiting for its frame to return. IBM claims that the use of multiple tokens traveling on the same network can boost efficiency to more than 95 percent on frames larger than 128 bytes.

Fiber-Distributed Data Interface (FDDI)

Network managers can go crazy agonizing over how to boost the speed of their networks because of the presence of not only 100-Mbps Ethernet but also asynchronous transfer mode (ATM), which is discussed in chapter 9. The quick-and-dirty solution today is still a fiber-distributed data interface (FDDI), the ANSI-based standard. FDDI provides 100-Mbps transmission between nodes, workstations, and hubs up to two kilometers.

In 1994, close to 30 vendors offered FDDI products, including bridges, routers, and gateways, as well as hubs and wire concentrators. FDDI-compliant products offer LED optical transcribers operating at a 1,300-nanometer light wavelength and stepped-index multimode fiber, with core and cladding diameters of 62.5 and 125 microns respectively. Fiber versions of FDDI are still very expensive; often the determining factors in choosing this technology are the distances at which nodes must be linked and security. The optical transmission over fiber makes data transmission virtually immune to interference from nearby machinery, as well as eavesdropping.

Copper versions of FDDI

Let's say that the major reason a network manager is looking at FDDI is because of network traffic congestion, but price is a major consideration. One alternative is to install a version of FDDI running over copper wire to gain 100-Mbps transmission speeds at a per-port cost of less than $1,000. Sound simple enough? Actually, here's where the fun really begins. There are several different copper versions of FDDI. The FDDI over-shielded twisted-pair wire, or SDDI specification, was published in May 1992. It uses the same 125-megabaud signalling rate as FDDI. Capable of supporting transmission up to 100 meters, SDDI uses a 1,700-millivolt electrical signal and a line-coding scheme known as nonreturn to zero, which is generally used only with IBM type 1 and type 2 shielded twisted-pair wire.

A year before the SDDI specification was published, several vendors, including Advanced Micro Devices, Chipcom, Digital Equipment Corporation, Motorola, and SynOptics, announced the "green book" STP specification. Using a transmission signal level of 700 to 1,200 millivolts, this specification uses a signaling rate of 62.5 megabaud, half the rate of FDDI and SDDI. Generally, IBM type 1 and type 2 cabling is used with a transmission distance of 100 meters.

Perhaps the version of copper FDDI that has generated the most attention is CDDI, a name trademarked by Crescendo Communications, now owned by Cisco Corporation. This version uses a transmission signal of 2,000 millivolts and a two-level coding method known as MLT2/3. The CDDI version transmits over either STP or EIA category 4 or 5 unshielded twisted-pair wire. The transmission distance varies between 75 meters for UTP and 100 meters for STP.

Clearly, network managers have to be very careful not to mix and match copper versions of FDDI. A standard that's not quite a standard means that interoperability isn't assured. Vendors claiming to support an ANSI standard for FDDI over copper wire usually means that they conform with the ANSI X3T9.5 committee's working (draft) documents. Even in 1995 there's the distinct possibility that reseller channels will still contain products that don't conform to any final set of specifications.

It's likely that all the wrangling over a final set of specifications for FDDI over-copper wiring combined with the announcement in 1993 that a 100-Mbps version of Ethernet was going to be shipped before the end of that year has pretty well slowed any marketing advances this technology was making. Crescendo, for example, saw drastic drops in sales of its CDDI product after Grand Junction announced fast Ethernet.

A second limitation to copper versions of FDDI and even FDDI itself is that it requires a different topology (dual ring) and different packet size than found in Ethernet. Because there's so much Ethernet in the installed base, network managers would prefer a solution that doesn't present new interoperability issues, such as the need for a new bridge or router or the acquisition of new network drivers. The birth of 100-Mbps Ethernet has probably relegated all versions of FDDI to niche markets. In 1994, prices for copper versions of FDDI cost approximately twice that of 100-Mbps network interface cards. The price of 100-Mbps Ethernet will continue to drop much faster than FDDI.

Developed by the American National Standards Institute (ANSI) committee X3T.9, the FDDI standard is a counter-rotating token-ring system capable of covering a very large area (200 Km) while transmitting data at 100-Mbps speed, a standard that ensures compatibility with IEEE 802.5 token-ring networks by maintaining the same frame fields found in that standard.

The FDDI model is reasonably consistent with the OSI model. As Figure 1.13 reveals, the FDDI physical layer is broken down into a physical layer protocol (PHY), which concerns itself with the actual encoding schemes for data, as well as a physical medium dependent layer (PMD), which provides the actual optical specifications. A media access control layer handles token-passing protocols as well as packet formation and addressing. Notice that there's also a set of station management standards to provide information on such tasks as removal and insertion of workstations, fault isolation and recovery, and collection of network statistical information. SMT uses the connection management protocol in conjunction with PHY line states to determine whether or not nodes entering the ring are linked together; unfortunately, SMT hasn't yet completed the lengthy process of being formally approved. I'll explain the effect of this lack of a standard on systems integrators later in this chapter.

FDDI was initially used primarily for "back-end" applications, such as connecting mainframe systems and mass storage devices, and for the backbone network function of connecting different networks. Today, it's ready to

OSI layer	FDDI layer
Data link	MAC
Physical layer	PHX (physical) PMD (physical medium dependent)

Figure 1.13 The structure of FDDI.

take its place along with token-bus and token-ring systems as a viable standard for large networks, while still performing the backbone function for Ethernet and token-ring networks.

What makes FDDI so appealing despite its expense is its speed of transmission and its dual-ring approach, which offers built-in protection against system failure. One major difference between IEEE 802.5 token-ring networks and an FDDI network is that a token-ring network circulates one token at a time on a 4-Mbps network and perhaps two or three tokens on a 16-Mbps network. On the 4-Mbps network version, a sending station transmits its token and then waits until the token is returned to it by the receiving station with an acknowledgment that the message has been received before passing the token to the next workstation on the ring. In an FDDI network, the workstation sending a message passes on the token immediately after transmitting the message frame and before receiving an acknowledgment that the message has been received. As a result, several message frames can be circulating around the ring at any given time.

Another difference between FDDI and token rings that enhances network speed is the use of a restricted token; it's possible to keep other workstations off the network while a time-crucial task is being performed. A timed token protocol ensures that low-priority messages won't clog up a network during peak hours. Timed token protocol uses both synchronous and asynchronous transmission. Workstations use a certain amount of transmission bandwidth defined for synchronous service, while the remaining bandwidth is used by workstations that transmit signals asynchronously when the token service arrives earlier than expected. They continue to do so until the expected time of token arrival, when they switch to synchronous transmission.

In addition to greater speed of transmission, FDDI networks enjoy a built-in redundancy that protects against system failure. The FDDI standard specifies a dual ring, one primary ring to carry information and a secondary ring to carry control signals. Figure 1.14 shows a typical FDDI network. Notice that, should a break in cabling result between stations A and B, it would be possible for them to continue to communicate through station C, which acts as a wiring concentrator. Note that it's possible to send data over both sets of cabling traveling in opposite directions so, if there's not a break in the cabling, a transmission speed of 200 Mbps is possible.

The basic components of an FDDI network

The FDDI standard defines the network components required including a single attached station (SAS), a dual attached station (DAS), and wiring concentrators. Single attached stations are attached to wiring concentrators using a star topology, as illustrated in Figure 1.14. Notice that the concentrators can include mainframes, minicomputers, and high-performance

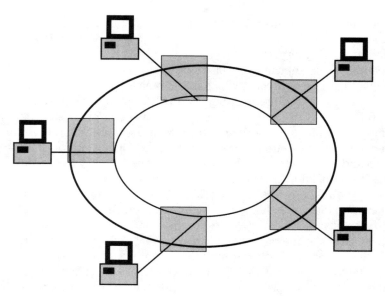

Figure 1.14 A typical FDDI network using dual cabling.

workstations. A cable break with a single attached workstation won't bring down the network because the concentrator is able to bypass the workstation and continue transmitting and receiving information.

These wiring concentrators can be very attractive for a systems integrator because they can connect anywhere from 4 to 16 workstations to the network at a much lower cost than using dual attached interfaces. Also, devices attached to concentrators can be switched off without affecting the network. Devices attached with a dual attached interface can have an adverse effect on an FDDI network if they're placed out of service because the FDDI network might assume the device is defective and try to remedy the situation by wrapping around itself, a phenomenon I'll discuss in more detail very shortly. Many industry experts expect FDDI network designs to use concentrators for PCs and other workstations, and more expensive but system-fault-tolerant dual attachment interfaces for minicomputer and mainframe links.

Double attached workstations on an FDDI network use dual cabling. The dual attached interface provides a system fault tolerance through its redundancy. In Figure 1.15, you can see an FDDI double attached network with a token traveling in one direction. A break in the cabling would cause the network to perform a function called *wrapping*, in which it activates the second ring to bypass and isolate the failed station. The network will continue to operate, but performance will decrease. Some vendors offer an optical bypass cable on their double attached interfaces so the connection between the right and left sides of the network can be maintained even with a cable break.

FDDI permits a maximum of 1,000 connections, with a maximum fiber-optic path length of 200 Km. The FDDI standard is 62.5/125-micron multi-mode fiber-optic cable with light generated from long-wavelength LEDs transmitting at 1,300 nm. Each station on an FDDI network functions as an active repeater, which helps explain why FDDI networks can be so large without signal degradation.

As touched on earlier in this chapter, another reason why FDDI networks are so fast is their use of a restricted token mode. Normally, FDDI networks use synchronous transmission for large blocks of information. Restricted to-ken mode reserves all asynchronous bandwidth for a dialogue between two workstations that want to use this type of transmission. All other workstations continue to broadcast in synchronous mode while this dialogue is taking place.

The FDDI frame

The FDDI frame, illustrated in Figure 1.16, can have a maximum size of 4,500 bytes, which makes it ideal for large file transfers. The preamble field synchronizes the frame with each station's clock using a signal that consists of 16 1-bits. Unlike an 802.5 network where the monitor station sets its master clock and all other stations use its transmission signal to set their own clocks, FDDI uses a distributed clocking approach, where each station's NIC sets its own clock. The starting delimiter field signifies the beginning of the frame and is followed by the frame control field. This field indicates whether the transmission is synchronous or asynchronous, what size address will be used (16-bit or 48-bit), and the type of frame to be found (an LLC frame or MAC control frame).

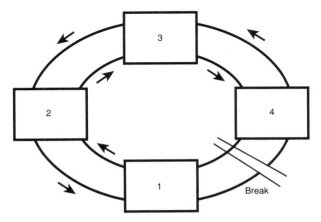

Figure 1.15 Wrapping after a cable break on an FDDI network.

Preamble	Starting delimiter	Frame control	Destination address	Source address	Data	Frame check status	Ending delimiter	Frame status

Figure 1.16 The FDDI frame format.

The destination and source address fields are self-explanatory. If the first bit in the destination field is set, then you have a group address indicating a broadcast message to all workstations on the ring. The data field is followed by a frame check sequence (FCS) field that uses a 32-bit CRC to check for errors. The ending delimiter (ED) field indicates the end of the frame, except for the frame status (FS) field, which a station uses to indicate it has detected an error.

Using multiple tokens on an FDDI network

Figure 1.17 illustrates the FDDI token. A station that wants to transmit seizes a token, absorbs it, and then adds its information to form a frame. This station begins transmitting and continues to do so until it runs out of information to send, or its token-holding timer expires.

The frames transmitted are repeated by other workstations around the ring. The destination workstation identifies its address on the frame and copies the information, checks for errors, and then sets a bit in the FS field if it detects an error before transmitting the frame. When the frame is returned to source workstation, it retransmits it if there was an error detected, or it purges the frame.

The moment a workstation releases a token, other workstations can seize it and begin transmitting frames. Workstations don't need to wait for the return of the frame they sent before releasing the token.

Integrating FDDI networks with existing LANs

The major approaches to linking FDDI nets with existing LANs are data-encapsulating bridges, translating bridges, spanning-tree bridges, and source-routing bridges.

The data encapsulation method used by companies such as Fibronics Inc. bundles data into FDDI format using proprietary algorithms. It takes a packet from a LAN and encapsulates it within an FDDI packet for its trip

Preamble	Starting delimiter	Frame control	Ending delimiter

Figure 1.17 The FDDI token.

around an FDDI ring. The encapsulating bridge at the receiving end strips the encapsulation from the packet and sends it along its way. The encapsulating and deencapsulating processes use proprietary algorithms that make this approach vendor-specific and incompatible with other vendors' data-encapsulating bridges, as well as other types of bridges.

The translating bridges approach offered by companies such as Fiber-Com Inc. readdress data using a protocol-independent method that's open rather than proprietary. A translating bridge takes a packet from a LAN such as Ethernet and translates the packet into FDDI packet protocol. At the destination end, a second bridge translates the FDDI protocol back into the original LAN protocol.

Fibre Channel

With support from the American National Standards Institute (ANSI), as well as both IBM and Hewlett-Packard, fibre channel is an emerging standard for high-speed transmission of data over fiber-optic cable at speeds exceeding 1 billion bits per second. Its major advantage beside speed is that it's a nonblocking switching technology, which means that multiple communications can take place without any collisions.

Another advantage of fibre channel is that it was designed to be compatible with several existing high-speed interfaces, including high-performance parallel interface (HIPPI) and small computer systems interface (SCSI). HIPPI is a parallel link for high-speed data running at 800 or 1,600 Mbps at distances of 25 meters or less, while SCSI is a low-cost interconnect between disks and personal computers. IBM successfully demonstrated a fibre channel network connecting 32 processors at a 1992 industry show (Supercomputing '92). Hewlett-Packard has developed a fibre channel interface card designed for workstations, supercomputer, and storage devices.

Even at this early stage of development, fibre channel is attractive compared to FDDI because it offers ten times the performance (one gigabit per second) at ten times the price. Another major advantage is that fibre channel defines three classes of service:

Connection service.   A dial-up point-to-point link.

Acknowledged connectionless service.   Electronic registered mail providing an acknowledgment of receipt.

Unacknowledged connectionless service.   Electronic registered mail that doesn't include acknowledgement of receipt.

It's no secret why IBM and Hewlett-Packard are behind fibre channel. Both companies have invested research funds in developing products using this technology for both the minicomputer and mainframe worlds.

A network manager, however, must consider some of the negatives of fibre channel before implementing it. Because this type of network is new to the world of PC LANs, there are no routers available to link existing Ethernet and token-ring LANs. Equally important is the question of whether fibre channel is a dead end that will be made obsolete by the development of asynchronous transfer mode (ATM), the switching technology discussed in chapter 9.

Computer Intelligence InfoCorp, a leading computer-industry market-research company, believes that the price of ATM adapter cards for personal computers will drop from 1993's $1,295–$1,595 range to under $700 by 1996. A network manager would be making a much safer bet in investing in ATM technology rather than fibre channel. Finally, in a fiber-based, multilink, "switched" topology, high-speed transmission channels are set up very much like telephone calls, which are switched through a telephone exchange. What this means is that fibre channel could take as long as 10 seconds to actually set up a high-speed connection. This time delay means that fibre channel could be used to link optical devices to network servers, but probably not for network links, which require real-time access.

AppleTalk on the Enterprise Network

So far we've been looking at a PC-centric enterprise network, but the corporate computing environment is also likely to include Macintosh computers, which use the AppleTalk network protocol. Because Macintosh terminology is so different from that found in the PC world, it's useful to begin our look at Macintosh networks by defining the basic building blocks. An AppleTalk local area network includes a Macintosh workstation, a hardware network interface, cabling, appropriate protocol software, and a network operating system.

While IBM PCs and compatibles equipped with AppleTalk network interface cards can operate on an AppleTalk network, we'll focus initially on an all Macintosh network.

Macintosh workstations

Macintoshes come with a built-in hardware interface for an AppleTalk network. If you want to use a Macintosh computer on an Ethernet or token-ring LAN, you'll have to add the appropriate network interface card. This AppleTalk interface is also found on Apple's LaserWriter printers.

LocalTalk

Here's where the terminology might get a bit confusing. The Macintosh's hardware interface contains low-level software responsible for transmission and media access control to Apple's LocalTalk cabling system, which uses shielded twisted-pair wire. In other words, the Macintosh's LocalTalk interface is capable of packaging bits into packets and then transmitting them, following the network bus's rules for media access at a speed of 230.4 Kbps.

Cabling

Apple's shielded twisted-pair cabling uses RS-422 signaling for transmission and reception over LocalTalk, and requires repeaters for distances greater than 1,000 feet. Two repeaters can extend the network to a maximum of 3,000 feet. The bandwidth of LocalTalk shielded twisted-pair cabling is limited, and likely to become saturated if a network grows much beyond 25 users. One solution is to use cabling with greater bandwidth, such as the coaxial cabling associated with Ethernet. A second solution is to use AppleTalk's ability to create a series of subnetworks and bridge them together.

CSMA/CA

LocalTalk uses a multidrop bus, the same scheme used by Ethernet. It also uses a media access method that's similar to Ethernet's CSMA/CD. Carrier sense multiple access with collision avoidance (CSMA/CA) is an approach in which a workstation that wants to transmit information first senses any activity on the network. If a collision occurs, all workstations avoid additional possible collisions by waiting for the network to be idle a specified amount of time plus an additional amount of random time that varies from station to station.

Unlike CSMA/CD, where hardware detects a collision, CSMA/CA works on the premise that a collision might have occurred. It relies on "handshakes" between sending and receiving workstations. When a workstation doesn't receive the appropriate "handshake" or control packet reply to its request to send information, it infers that a collision has occurred and begins the process of requesting the right to transmit all over again. Let's examine this process in more detail by looking specifically at the LocalTalk protocol responsible for this data link access method.

LocalTalk's Link Access Protocol (LLAP)

LocalTalk has its own link access protocol known as the LocalTalk link access protocol (LLAP). Figure 1.18 illustrates the LLAP packet. A preamble indicating the start of a frame is followed by destination and source node identifiers. These 8-bit addresses can range from 1 to 127 for user nodes and 128 to 254 for server nodes under AppleTalk's first version, known as Phase 1. The number 255 is reserved for a broadcast node ID. When a workstation starts up, it randomly assigns itself an ID and then tests to see if another node is already using it by sending out an inquiry control packet.

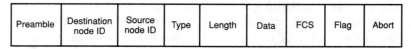

Preamble	Destination node ID	Source node ID	Type	Length	Data	FCS	Flag	Abort

Figure 1.18   The LLAP frame.

The LLAP type field indicates whether a packet is carrying control information or actual data. The data length field indicates the amount of data the packet carries; an LLAP packet can carry between 2 and 600 bytes of data. A frame check sequence field is used for error checking and is followed by a trailer flag (01111110) field and an abort field that indicate the end of a frame. LLAP uses a technique known as *bit stuffing* to ensure that no other field contains more than five consecutive 1-bits. It will insert a 0-bit after five consecutive 1-bits to ensure the flag field's uniqueness. A receiving workstation's LLAP will reinsert the appropriate 1-bit so the data can be read correctly.

How workstations transmit using LLAP

When a workstation using LLAP wants to transmit, it checks the network until it has been idle for at least 400 microseconds and waits an additional random period. The source workstation then sends a request-to-send packet to the destination workstation, which replies with a clear-to-send packet. Then the source workstation transmits its data packet. If the source node doesn't receive a clear-to-send control packet, it assumes there has been a collision and once again waits until the network has been idle for at least 400 microseconds.

Broadcast transmissions work somewhat differently. A source workstation waits at least 400 microseconds and an additional period of time before sending out a request-to-send control packet with a broadcast address. It then checks the network for a maximum period of 200 microseconds before broadcasting its transmission. The source workstation will attempt up to 32 retransmissions before reporting failure.

AppleTalk Phase 1

AppleTalk refers to the entire suite of protocols that comprise Apple's own layered network architecture. AppleTalk Phase 1 imposed the limitation, under LocalTalk, of no more than 32 nodes in a single network, including workstations, servers, and peripherals. It was possible, however, for several AppleTalk networks, each one limited to a maximum of 32 nodes, to be bridged together.

Apple uses the term *internet* to define two or more local area networks linked together with a router or gateway. When a router combines network segments into an internet, as shown in Figure 1.19, the networks remain independent of each other. Networks connected together by routers are known under Apple's terminology as *zones*. A zone consists of logical grouping of networks on an internet; it's very important to realize that a zone need not be identical to the original LAN's physical configuration. In other words, nodes that share functionality (accounting, word processing, etc.) might be grouped together in a zone by a network manager even though they're located on different LANs at the same physical site that comprise the internet. Often, a backbone network (also shown in Figure 1.19) is used to reduce traffic congestion between networks on an internet. The name of the game is to reduce the number of routers necessary to transmit information from one node to another.

In addition to its own LocalTalk, Apple's AppleTalk Phase 1 provided drivers for Ethernet (EtherTalk). This meant that it was possible to design a Macintosh network using coaxial cabling and Ethernet NICs to achieve 10-Mbps transmission speed. The 8-bit field AppleTalk assigned to node addressing meant that, even under Ethernet, there was a limitation of 254 network nodes, including workstations, printers, and modems. For some corporations, this restriction simply was not acceptable.

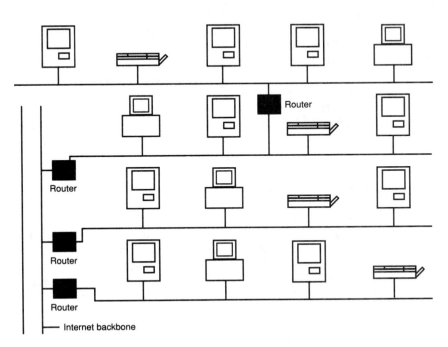

Figure 1.19 An AppleTalk internet.

AppleTalk Phase 2

AppleTalk Phase 2 added improved internet routing, for up to 1,024 inter-connected AppleTalk networks, and it provided extended resource grouping to support up to 256 zones per network. Even more appealing to systems integrators concerned with connectivity, it provided drivers for token rings (TokenTalk) and enhanced drivers for Ethernet.

Perhaps the most significant change under Phase 2 was replacing Phase 1's eight-bit address field, which limited AppleTalk workstations to addressing a maximum of 254 nodes, with a 24-bit field that provided the theoretical ability to address more than 16 million different nodes (2^{24}). One serious problem with this change is that Phase 1 packets are invisible to Phase 2 nodes because of the incompatibility of the addressing schemes. While Phase 2 can handle the addresses of newer Apple Laserwriters, including the SC, NT, and NTX, it can't handle the addresses of the older LaserWriter and LaserWriter Plus models. Also, some software (particularly network management software offered by third-party vendors) might work fine on Phase 1 networks but fail under Phase 2.

Moving to Phase 2 means that companies who want to use Ethernet need to upgrade the EtherTalk drivers on each Macintosh node. Much more seriously, however, Phase 2 requires upgrading all hardware and software network routers. Routing is completely different under Phase 2.

For a network manager, there are limited options. One choice (Apple's suggestion) is to bring down the entire AppleTalk network and upgrade all nodes to Phase 2. This expensive option might increase your Apple stock, but it might cause your company stock to decline rapidly. A second option is to divide the AppleTalk network into several small internets and then selectively upgrade each one. This option will disrupt internet traffic, but carefully designing the internets could minimize this problem. The third option is the equivalent of a dual-protocol stack. Routers could be outfitted with both sets of drivers. This approach would work, but it would also cause serious degradation in network efficiency because of the imposed overhead.

As I pointed out in my earlier discussion of AppleTalk Phase 2, AppleTalk has link access protocols for token-ring (TokenTalk) and Ethernet (EtherTalk) networks as well as its own LocalTalk hardware. The drivers that AppleTalk provides are able to handle the different addressing schemes, different-sized packets, and different media-access methods required by the different types of networks. Figure 1.20 illustrates how an EtherTalk packet differs from the LocalTalk packet discussed earlier in this chapter. Notice that it contains the Ethernet-specific fields required for communications on an Ethernet network.

Ethernet destination	Ethernet source	Type	AppleTalk source	AppleTalk type	Length	Pad

Figure 1.20 An Ethertalk packet.

Arcnet LANs on an Enterprise Network

Slow, difficult to link to other topologies, but virtually indestructible, attached resource computer network (Arcnet) is one of the oldest LAN technologies, developed by Datapoint in 1977. Major vendors that license and support Arcnet include Standard Microsystems Corporation (SMC), Acer Technologies, Thomas Conrad, and Allen Bradley. Arcnet is found in more than one million active nodes, yet it never seems to garner the respect or attention of the industry. After almost two decades of life as a nonstandard, the topology has finally become an ANSI standard. This status, though, has probably come too late to breathe new life into Arcnet sales. Arcnet is dying, but its installed base is so large that it must be considered a factor in the construction of any enterprise network. Let's see what has caused Arcnet to grow so large.

Why Arcnet has been successful

Arcnet has been popular with small as well as large systems integrators because it's very inexpensive; an Arcnet card might cost ⅓ the price of a token-ring NIC. It can use a single shielded untwisted-pair cable, which most companies can spare, and it runs on RG 62/U coaxial cabling, the same cabling many companies have for their 3270 terminals. IBM shops that replace their 3270 with PCs can use the same coaxial cabling to build an Arcnet LAN. Companies that do need fiber-optic cabling can also use this medium with Arcnet.

Besides being one of the least expensive and most reliable LAN technologies, Arcnet is also one of the most flexible networks. Adding and removing nodes, initializing the network when first powered on, and recovering lost tokens are functions all handled by Arcnet hardware. It's a joy to install compared to other LAN hardware, including token ring and Ethernet, and it's much easier to diagnose problems. As you'll observe in a few moments, its topology is probably the most flexible found in the industry. It's very important to keep in mind that Arcnet hardware can run virtually any NetBIOS-compatible network operating system. You'll find LAN Manager and IBM's PC LAN program running on Arcnet, but the most popular choice for larger businesses is the Arcnet/NetWare combination.

Topology

Arcnet supports bus, star, and distributed star topologies. Each segment of an Arcnet bus can contain up to eight nodes, daisy-chained together, and can extend to 1,000 feet. Adding an active link to a bus segment will extend the segment's range another 1,000 feet, point to point, as revealed in Figure 1.21.

Arcnet's star topology is popular because it's the easiest arrangement to troubleshoot; also, the failure of a single workstation won't bring down the entire network since each workstation has its own cabling connection with a hub. Up to eight workstations can be connected to a central hub, with a 2,000-foot maximum. Passive hubs, on the other hand, are limited to distances of 100 feet. Active hubs can be connected together to build a network with a maximum distance of four miles. Figure 1.22 illustrates a typical Arcnet star topology with both an active and a passive hub.

I mentioned earlier that Arcnet's topology is extremely flexible. It's easy to connect bus segments with an Arcnet star or distributed star. Two port repeaters can connect the bus segment on one side with a star hub on the other side.

Arcnet's access method

Arcnet uses a token bus access method. It's a physical star or bus, but a logical ring. It's a noncontention network in which each workstation has a turn to transmit based on its NIC address, which is set using an eight-position DIP switch. Each NIC knows its own network address as well as the address of the node to which it will pass the token. The highest-addressed node closes the logical ring by passing the token to the lowest-addressed node.

Figure 1.23 illustrates Arcnet's token-passing network functions. When node 150 has completed its network token time, it passes the token to node 10 and the process around the logical ring begins all over again. Arcnet's token bus technology was developed in a minicomputer transaction-oriented environment, where short network bursts consisting of re-

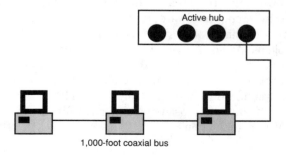

Figure 1.21 An Arcnet bus topology with an active link.

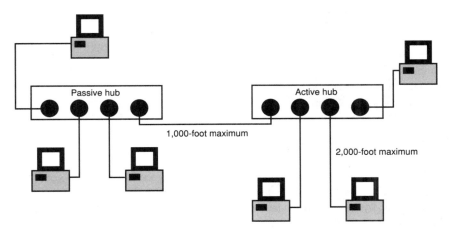

Figure 1.22   An Arcnet star with active and passive hubs.

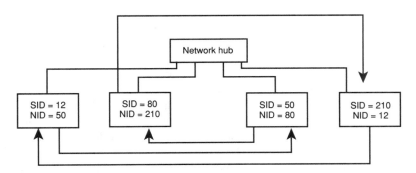

Figure 1.23   Arcnet's token bus approach.

quests for database information place a premium on moving relatively small data packets quickly and efficiently. Its low overhead and efficient noncontention token bus transmission at 2.5 Mbps makes it competitive with Ethernet under heavy traffic conditions, even though it has substantially less bandwidth.

The Arcnet packet

The major reason Arcnet is incompatible with IEEE standards is its addressing. It uses an eight-bit, locally administered, station-address format, while IEEE 802 LANs use a 48-bit, globally administered, station-address format. There are simply too many active Arcnet nodes to change the addressing scheme. Instead, Arcnet vendors have fought a long political battle to persuade an ANSI committee to approve an Arcnet standard.

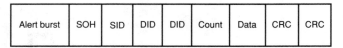

Alert burst = Six 1-bits identify a packet
    SOH = Start of header (1 byte)
    SID = Source workstation ID (1 byte)
    DID = Destination workstation ID (2 bytes)
    Data = 1-508 bytes
    CRC = Error checking (2 bytes)

Figure 1.24   The Arcnet packet format.

Figure 1.24 describes the Arcnet packet format. Alert consists of six consecutive bursts that identify a packet. Note that the source and destination IDs are one byte and two bytes, respectively, far different than the IEEE Ethernet and token-ring formats described earlier in the book. Note also how relatively small the data field is; it's dwarfed by Ethernet's approximately 1,500 bytes. When Ethernet has to retransmit its packets because of frequent collisions, the beauty and simplicity of Arcnet becomes apparent.

Arcnet's age is apparent, though, in the way nodes communicate with each other, using a character-oriented protocol. If node 20 wants to send a packet to node 40, it first sends a free buffer enquiry (FBE), which asks if node 40 is ready to receive a transmission. Node 40 responds with an acknowledgment (ACK), or a negative acknowledgement (NAK) if it declines the FBE. When node 40 receives the data packet from node 20, it checks the CRC to ensure there was no error in transmission and then transmits an ACK to node 20 indicating that everything arrived successfully. If it fails to send an ACK, node 20 determines that there must have been an error in transmission. Assuming the message did arrive correctly and an ACK was transmitted, node 20 then issues an FBE to the next node scheduled to use the token.

Arcnet Plus at 20 Mbps

One of the most recent developments has been the announcement of a 20-Mbps Arcnet Plus network. Datapoint has the proprietary rights to this technology, but has licensed SMC and NCR to sell products. Arcnet Plus is able to dynamically vary its data signaling rate so it's backward compatible and interoperable with standard Arcnet. Network supervisors need not pull any existing Arcnet NICs or cabling, but can introduce Arcnet Plus network segments selectively, such as on backbones, to optimize network performance.

Arcnet Plus supports packet sizes from 12.5 to 4,224 bytes, compared to a maximum size of 516 bytes with standard Arcnet. The network supports up to 2,047 nodes per segment over standard coaxial or twisted-pair wire, compared to standard Arcnet's 255 nodes. This enhanced version of Arcnet will support IEEE 802.2 48-bit addressing as well as the 802.2 media access control layer. Arcnet Plus will enhance compatibility with other 802 networks, and also continue to support traditional Arcnet addressing.

Network design when mixing the two Arcnet technologies might be a headache for systems integrators, however. Standard Arcnet NICs can't recognize addresses above 255, so it's important to minimize communications between these two sides of a mixed network. Current hubs can transmit only at 2.5 Mbps, and it might be a while before new hubs reach the market. My advice is to wave a fond farewell to Arcnet and stick with the much more interoperable Ethernet and 10BaseT systems.

Client/Server Meets Peer-to-Peer NOS on the Enterprise Network

Mixed topologies are just one example of the growing complexity of enterprise networks. Another reason why enterprise networks are becoming even more difficult to manage is the growing coexistence of client/server and peer-to-peer network operating systems.

Artisoft was the first company to capitalize on this phenomenon, with its LANtastic for NetWare product. The company enjoyed particular success among Fortune 1000 companies where departmental managers ran the peer-to-peer NOS on top of their NetWare network operating system to provide greater control of local resources. A departmental secretary might have a Hewlett-Packard LaserJet III as a local printer, but use it only 20 percent of the time. By linking all departmental users to the printer and to each other via LANtastic for NetWare, the departmental manager is able to use local resources more efficiently; department employees need only walk down the hall to the large printer accessible via the company's NetWare LAN when a long document or a document filled with graphics needs to be printed. Other jobs can be routinely printed on the HP III using the peer-to-peer software and existing cabling for the NetWare LAN.

Microsoft's Windows for Workgroups has progressed from its first anemic version to a powerful peer-to-peer NOS that runs reasonably well with NetWare, having overcome its early interoperability problems. Artisoft has gone a step further by licensing Novell's network core protocol (NCP) as well as the NetWare 4x engine. Now an Artisoft LANtastic peer-to-peer LAN can read files on a NetWare server. The convergence of client/server and peer-to-peer network operating systems continues with the indication from IBM that it will incorporate peer-to-peer capabilities in OS/2, with a client/server network operating system LAN server running on top of it. Similarly, Novell's Personal NetWare now contains peer-to-peer functionality. If used in conjunction with NetWare, users would have both client/server and peer-to-peer functionality.

The coexistence of peer-to-peer and client/server network operating systems presents a major problem for network managers. How can functions performed under both systems be managed? What about security? A peer-to-peer NOS exposes individual users' hard disk drive directories to use and perusal by other network users. Finally, who is responsible for managing peer-to-peer networks running on top of but invisible to client/server LANs? Because the price of peer-to-peer software often costs less than $100 per node, departmental managers can purchase it from their own budgets, sometimes without the knowledge of network managers. The amount of memory required by these peer-to-peer programs means that you need a fairly sophisticated knowledge of memory management to make sure the right programs load into high memory and that memory conflicts are avoided. Despite these problems, departmental managers persist in bringing peer-to-peer programs in by "the back door."

The struggle for control between these departmental managers and the network manager are reminiscent of the struggle a few years ago between the MIS manager and the departmental managers who often brought in personal computers to solve problems so they wouldn't have to wait for MIS support and funds.

Printing in an Enterprise Network Environment

Besides the problems of mixed topologies and mixed network operating systems, network managers in enterprise network environments must also grapple with the problems associated with printing in such complex environments. One problem associated with such a network is the need to support both PostScript and print control language (PCL). Some printers, such as those offered by QMS, are smart enough to sense the type of job being transmitted and select the appropriate page description language to handle it.

Network managers need to test their real-life applications on a network printer prior to purchasing it. Often, vendors who sell more expensive network printers that contain network interfaces for direct connection to a LAN will permit a network manager to evaluate the printer for a brief period of time. If this is impossible, then make a list of all major applications and get the vendor's assurance that these applications will run acceptably on the printer.

Managing network printers

One of the most distressing parts of the network manager's job is printer management. NetWare users frequently send a print job to a network print queue without knowing if the job is actually printed. Network managers are frequently unaware of printer problems until a print queue contains several jobs waiting to be printed. Other printer information that would be valuable for network managers to receive includes notification of toner status, paper status, paper jams, and general failure.

Up until very recently printer vendors offered only proprietary solutions to network printer management. Hewlett-Packard, for example, offered users of its Jet Direct cards on NetWare LANs the ability to receive information such as whether or not a printer was out of paper. Unfortunately, most large corporations have a large number of different printers and being able to manage only a select group is unacceptable to most network managers. The solution has to be an industry-wide set of specifications.

The network printing alliance protocol (NPAP) is a set of specifications being developed as an industry-wide standard for late 1994 completion. Companies sponsoring this standard include Lexmark, QMS, Kyocera, NEC, and Okidata. Information from a printer will travel across networks, via serial links or even via bidirectional communications from a parallel port. Significantly missing from this list is Hewlett-Packard. It clings to its proprietary printer job language (PJL). While not as elaborate as NPAP, it does give Hewlett-Packard a competitive advantage until NPAP reaches fruition.

NPAP is a packet protocol that enables a computer to send commands to the printer and also the printer to respond. The host computer inquires about a printer's basic characteristics, such as its model name, serial number, resolution, emulations, and number of fonts available under each emulation. The host uses the information it receives to update its configuration tables, verify features and fonts, and conform the availability of printer supplies. Once all this information is received and verified, the host computer selects its emulation mode and begins sending data to the printer.

The start-of-packet byte is always the equivalent of a decimal 165 in order for the printer and computer to become synchronized. Because NPAP is still in its infancy, only six base commands and 32 subcommands have been defined under level 1. The command byte can theoretically contain up to 240 base commands, so NPAP has plenty of flexibility to become very sophisticated in the future. NPAP will eventually work with the desktop management task force (DMTF), a standard management interface for desktop computers first developed in 1992. This standard is discussed in chapter 11. There are many advantages to a desktop standard that encompasses printer information as well as computer information.

Summary

Ethernet was the first nonproprietary local area network, and it still commands the largest number of users. 10BaseT, the IEEE 802.3 standard for unshielded twisted-pair wire transmission at 10 Mbps, is the most popular version of this topology. The 100BaseX and 100BaseVG versions of Ethernet offer 100-Mbps performance to eliminate some of the traffic bottlenecks now found on large Ethernet LANs. The IEEE 802.5 standard describes a LAN using a token-ring, noncontention approach. The vast majority of customers now purchase 4/16-Mbps switch-selectable network interface cards for token-ring networks.

The fiber-distributed data interface (FDDI) is a 100-Mbps transmission-speed fiber-optic network modeled on the IEEE 802.5 standard, but it has some significant differences. It uses a completely different error-correcting scheme that features a double ring and a wrapping technique. Network timing is handled by each workstation rather than by 802.5's monitor station. Rather than the token reservation system used by 802.5 networks, FDDI uses a timed-token system. A twisted-pair version of FDDI has proven popular. It offers cost savings, but not the distance and security features found with the fiber-optic version.

Enterprise networks might also contain some AppleTalk and Arcnet LANs. Companies with larger AppleTalk LANs are replacing the slow LocalTalk with EtherTalk or TokenTalk, while Arcnet is being replaced with Ethernet. A 20-Mbps version of Arcnet has come too late to help revive this aging topology despite its reliable performance.

2

Unlocking the
Mystery of Protocols

In this chapter, you'll examine:

- Network protocol—what it is and how it works

- The open systems interconnect (OSI) model, as well as government OSI profile (GOSIP)

- The significance of transmission control protocol/internet protocol (TCP/IP)

- How companies will migrate from TCP/IP to OSI

- Xerox network system (XNS) protocol

Take a deep breath before starting this chapter and keep telling yourself how good this material is for you. After ten years' experience working with LANs, I've concluded that, while protocols might not be the secret of life, they do provide the key to understanding internet connectivity.

In the novel *Shogun*, an Englishman in sixteenth-century Japan is led through a maze of paths surrounding a castle. At each gate, his companions present a pass that's carefully examined before they're allowed to continue. The Englishman cannot read or understand the passes, but he knows that if anything is wrong with any of the passes he won't be able to complete his journey. In a similar way, data on a network is packaged in just the right order, along with accompanying control information. The data and control information is placed in the equivalent of an envelope that must be addressed correctly. Unless everything is done completely according to the rules (or *protocol* in data communications terminology), the data won't reach its destination.

The focus in this book is when the data needs to travel not just to another workstation on the same network, but to a workstation on a completely different network. Here's where protocol really becomes crucial. Data and control information that can travel easily around its own workstation's network, with its current packaging and envelope, must carry additional information (in effect, another pass or envelope) to enter and travel on another network with a different protocol.

Many networks built around minicomputers have adopted the transmission control protocol/internet protocol (TCP/IP), while many microcomputer networks turned originally in the direction of the Xerox network systems (XNS) protocol and later toward IBM's NetBIOS interface. The open systems interconnect (OSI) model with its suite of protocols that are international standards continues to grow in popularity, albeit very slowly.

In this chapter, we'll examine these different protocols and look at how they package data and control information. We'll also look at the differences between the current dominant internetwork protocol (TCP/IP) and the OSI model that will eventually replace it. You can count on referring to this chapter later when you examine the bridges, routers, and gateways that link different networks together by dealing with these protocol issues.

What Are Protocols and Why Are They Important?

While an operating system manages a computer's resources and a network operating system manages an entire network's hardware and software resources, there can be no network communication without protocol. *Protocol* is the set of rules that govern how packets containing data and control information are assembled at a source workstation for their transmission across the network, and then disassembled when they reach the destination workstation. Assume that you just received a message from a network user on another planet. This message contains a series of characters and numbers with no spaces or apparent punctuation.

You can decode an English sentence because we know the rules, or protocol, of the language. You know that sentences begin with capital letters and end with periods. You know that individual words are separated by spaces. You also know that proper nouns are capitalized and that verbs often end in "ed." Without knowing the protocol that a message follows, you would have no idea which part of the message contained information for you and which part was, say, the address.

Protocol permits network communication by specifying which bits represent a greeting, which bits represent error checking, and even which bits indicate the size of the packet's data portion. Without this information, a receiving workstation would have no idea how to decipher a message.

A basic data communications model

In order to discuss different protocols, we need a basis of comparison, a set of terms to describe the common elements in any network communication. Some basic data communications terminology has evolved over the years and has become an integral part in defining the standards. An *application process (AP)* is software such as an inventory program or an accounts receivable program. It resides on a device described as *data terminal equipment (DTE)*, which could be a computer or terminal. If the application process needs information from a second computer, it needs to establish communications with another DTE.

To establish communications with a second DTE, the computer uses *data circuit terminating equipment (DCE)*, such as a modem, to provide an interface from the DTE to the communications network, which can take the form of a cable or a digital or analog phone line. The information is transmitted over this communications network to a second DCE, which converts the information into a form that the application process running on the second DTE can understand. Figure 2.1 illustrates the elements found in a basic data communications model.

Protocols in a data communications model

The key to standardizing the interfaces found in the data communications model is the development of *protocols*, which I'll define more fully here as sets of very specific rules on how data communications devices exchange information. Protocols cover such issues as how data is transmitted (what pin or wire carries what kind of signal), what kind of error checking takes place, and how data should be assembled with accompanying control infor-

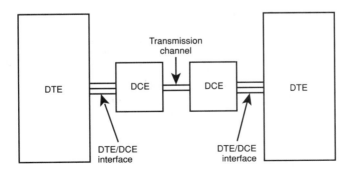

Figure 2.1 The elements of a data communications model.

mation into packets for transmission, and then disassembled successfully at the receiving end.

Protocols cover virtually all phases of communications including the synchronization of the receiving and sending computers' clocks, the method of coding the binary data into corresponding voltages, and even instructions on how information can be routed across several different networks with different addressing schemes without losing its integrity.

Moving Toward Standards

There are a number of U.S. and international organizations that have been working for several years to develop a set of standards for data communications equipment in general and local area networks in particular. The benefits of standardization include a reduction in equipment costs and the ability to link different devices, and mix and match hardware and software products from different vendors.

How standards are created

A set of protocols defining a particular data communications model or interface must undergo several different steps before it can become an international standard. In the case of the International Standards Organization (ISO), the protocol starts as a WP (working paper); then it becomes a DP (draft proposal). The next stage, DIS (draft international standard), is followed by the status of IS (international standard).

In the discussion that follows on the open systems interconnect (OSI) model, note that much work still remains in the international standard adoption arena. While most hardware standards have been adopted with little fanfare, software standards have been much slower to evolve; unfortunately, it's precisely these software standards that will really have a profound effect on local area networks.

The Open Systems Interconnect Model

You've just observed how important protocols are for ensuring effective communications by defining rules to be followed. In the field of data communications, protocols are very complex because of the amount of information that must be agreed upon for an interface to be effective. Since technology is constantly changing, after several years of discussion and negotiation among committee members with different viewpoints, often an interface or protocol will have been replaced by a new, vastly improved technique.

Layered protocols

Most standards organizations use a layered approach for developing protocols, where each layer concerns itself with some very specific functions and services. Since these standards must cover a variety of different possible hardware and software configurations, often several different layers of protocols are required to define a particularly complex communications interface. The advantage of this approach is that changes can be incorporated in one layer without having to redevelop the entire model. Since each layer is developed by a different subcommittee, it tends to overlap some of the functions and services of other layers. The result is the one major disadvantage of layered protocols—a significant amount of overhead. It takes more bits to cover all the functions and services defined in each layer than it would if two computers were communicating using a manufacturer's own proprietary operating system, where there are fewer options and defaults to be represented.

The rationale for the OSI model

The International Organization for Standardization (ISO) in conjunction with the Consultative Committee on International Telegraphy and Telephony (CCITT) developed a layered set of protocols known as the open systems interconnect (OSI) model to facilitate communications between computer networks. One of the major goals of the OSI model is that in the not-too-distant future it will be relatively easy for computers using OSI-compliant hardware and software to exchange information regardless of the fact that they were manufactured by different vendors. End users will be freed of the compatibility worries that still characterize systems with multi-vendor equipment.

Figure 2.2 illustrates how the layered protocols composing the OSI model facilitate the transfer of information from one computer to another. Note that each layer, with the exception of the physical layer, adds a header containing control information for its corresponding counterpart on the other computer. The data link layer even adds a trailer with additional control information. The control information found in the headers and trailers contain such key negotiable data as the form the information will take (will it contain floating-point numbers?), the address of the sending and receiving workstations, the mode of transmission (full duplex, half-duplex, etc.), the method of coding the information (EBCDIC, ASCII, etc.), and the type of error checking that will take place. After this information is received by the second computer as a bit stream, it's reassembled into frames and the control information is stripped off by each corresponding layer as the frame

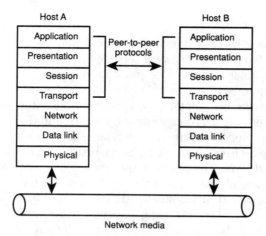

Figure 2.2   The OSI model.

moves up through the layered protocols until only the original data remains for the application program.

Layers in the OSI model

It might be helpful to view the individual layers of the OSI model as groups of programs designed to perform specific functions. One layer, for example, might be responsible for providing data conversion from ASCII to EBCDIC and have the necessary programs required to perform this task. The programs might contain individual modules, which are known in the OSI model as *entities*. Each layer provides services for the layer above it while requesting services from the layer immediately below it.

The upper layers request services in a rather general fashion; they might want some data routed from network A to network B. The actual mechanics of how the data needs to be addressed to ensure that it's routed correctly are left to the lower-level layers that provide the services. The communication between layers involves different types of transactions, known as *primitives*.

Primitives

These primitives involving transactions can take several forms, including requests, indications, responses, and conforms. A layer serving as a service user can invoke a function by requesting an action, such as data encryption. The layer serving the role of service provider will issue a confirmation indicating that the function has been completed ("yes, the data has been encrypted"). Sometimes the request is for action to be taken by a layer on a second computer. This request is received by the layer on the second computer as an indication primitive. The layer replies by issuing a response primitive that informs the layer on the first computer that the requested function has been completed.

It might be helpful to think of these primitives as control information that takes the form of certain bit patterns in the frames that are transmitted during the data communications process. The way the OSI model facilitates communications between data networks using this system of primitives to relay control information is directly applicable to how workstations communicate on local area networks.

The application layer

Under the OSI model, an application program that needs to accomplish a specific task, such as updating a database on computer B, sends specific data in the form of a datagram to the application layer. One of the prime responsibilities of this layer is to determine how the application program's request should be treated, in other words, what form the request should take. If the application program requests remote job entry, for example, this might require several programs to collect the information, organize it, process it, and then send it to the appropriate destination. Electronic mail is also a major function for the application layer. In chapter 11 you'll examine some of the X.400 OSI protocols for electronic mail and X.500 protocols for a global directory, in conjunction with the chapter's description of how electronic mail is moving toward global communications and breaking down the barriers between heterogeneous networks.

Application service elements in the application layer.   The application layer also contains several application common service elements (ACSE) and specific application service elements (SASE). The ACSE services are available to application processes on all systems. They include such services as quality-of-service parameters.

A workstation on a local area network in Los Angeles, for example, might need to communicate over a wide area network via modem with a mainframe computer in Boston. Because the quality of the phone line might prove unsatisfactory, the application process running on the LAN might request a quality of service that includes an acknowledgment that all information was received and understood. This quality of service is like a request at the post office that your package's delivery be acknowledged with a return receipt.

The specific application service elements (SASE) provide services for specific applications such as file transfer and terminal emulation. If an application program requires a file transfer, for example, one key application layer protocol that would be involved is file transfer, access, and management (FTAM).

Imagine, if you will, the implications for the future when local area networks as well as mainframe computers run OSI-compatible software. Since FTAM functions as a virtual filestore, with its own directory service, programs will be able to access databases unaware of the file's actual location. Since FTAM supports a variety of different types of file structures, including sequential, ordered hierarchical, and general hierarchical, information will be able to flow from one database located on a distant Unisys computer, update a second database running on a local area network in Los Angeles, and finally be updated by a third database residing on an IBM mainframe in Phoenix.

Another major SASE found in the application layer is virtual terminal (VT) service. VT is a sophisticated service that frees a computer from having to send the appropriate signals to address all terminals residing on a second computer. The first computer can use a set of virtual terminal parameters and leave specific terminal configuration concerns to the second computer, which has already been configured to address the signaling requirements of its own terminals.

Other SASEs in various stages of development include transaction processing, electronic data interchange (EDI), and job transfer and manipulation (JTM). The development of an OSI standard for EDI, in particular, could be very significant for LAN users. Workstations on an LAN could create purchase orders, for example, and then transmit the information electronically directly to the manufacturer or distributor, where the data would automatically go into an invoice. Inventory could be checked and decremented automatically and arrangements made to ship the goods all without a significant amount of paperwork or delay.

Network management functions in the application layer. Network management has become an issue as data networks have become more complex. As voice and data become integrated parts of data communications and local area networks link more and more with wide area networks and mainframe computers, there's an increased need for an effective method of managing and controlling this information. IBM has offered its NetView and NetView/PC as the solution, while Hewlett-Packard has countered with its OpenView software package.

The problem up to now has been that there have been several proprietary solutions but no effective international standard for network management. Under the application layer of the OSI model, there have been several management information protocol specifications working their way toward the position of international standard. I'll discuss the movement toward international standards for network management in chapter 11.

The presentation layer

The presentation layer concerns itself with how information is physically displayed or presented. Data might need to be processed as floating-point numbers, and certain database fields might require both characters and numbers. Some fields might even require graphics.

The presentation layer facilitates communication by ensuring that application processes that want to exchange information can resolve any syntax differences that exist. The two processes must share a common representation or language for communication to take place.

The presentation layer's real value to the other OSI layers is its creation of abstract syntax notation (ASN.1), which is its way of describing a file's data structures. ASN.1 is used by the application layer for all its file transfers and virtual terminal information. It's also essential for the encryption function, a major concern for managing large networks, that can be done at the presentation layer. The development of an OSI standard at this level will have a significant impact on facilitating machine-to-machine communications.

The session layer

Think of the session layer as an administrative assistant with a good eye for detail who is responsible for arranging all the details for an important upcoming meeting between two executives. In reality, the session layer's activities facilitate the session, resulting in an interchange of information between two application processes, but the analogy holds together rather well.

The session layer is concerned about such basic issues as the mode of transmission and synchronizing points. In other words, will transmission between the two application processes be half-duplex (the processes alternating sending and replying), or full-duplex (both processes sending and receiving at the same time)? In half-duplex transmission, the session layer provides the side that speaks first with a data token. When it's time for the other side to respond, it receives the data token. Only the side with this data token is permitted by the session layer to speak.

Similarly, synchronizing points represent points during the "conversation" that the session layer checks to ensure that actual communication is taking place. If you've ever observed two Japanese businessmen carrying on a conversation, you will certainly remember that both were nodding their heads and saying "Hi." This doesn't mean the two businessmen agreed with each other; it simply means they were indicating that they heard and understood what the other person was saying, since "hi" means "yes" in Japanese.

Another concern of the session layer is how to reestablish communications if they're disrupted during a session. A synchronization point between pages of a text, for example, would be a logical approach since reestablishment of a session wouldn't require beginning again and repeating all text already received correctly.

In a similar vein, this layer is also responsible for the details associated with an orderly (known as a "graceful") release of the connection when a session ends. There are also procedures for what is known as an "abrupt" release of a connection, a situation where one side ends communications and refuses to receive additional data from that moment on.

The session layer doesn't automatically accept every connection. It might issue a refuse connection primitive if, for example, it determines that it would result in too much network congestion or because the application process requested is unavailable.

The transport layer

The transport layer is of great importance to computer network users because it provides the quality of service required by the network layer. An analogy that might help clarify the transport layer's responsibilities is to think of it as a collection of special services available for additional cost at your local post office. In other words, for an additional fee you can receive a return receipt indicating that a letter was delivered to the person you specified. Similarly, you might require expedited service, such as specifying that a package reaches Boston by the next day. While the U.S. Postal Service collects fees for these high-quality added services, the price paid by a computer user running OSI-compatible hardware and software is the overhead of additional bits required to provide information on the status of these possible services.

There are three different types of network service provided by the transport layer. Its type A service provides network connections with acceptable residual error rates and acceptable rates of signalled failures. Type B offers an acceptable residual error rate and an unacceptable rate of signalled failures, and type C provides network connections with residual error rates that aren't acceptable to the session layer.

Why have classes of service that provide unacceptable error rates? The answer is that many network connections use additional protocols that provide the required error detection and recovery, and the required overhead for this service under the transport layer isn't necessary.

On the other hand, the transport layer does offer programmers the opportunity to write programs for the application layer for a wide variety of networks without concerning themselves with whether or not transmission on these networks is reliable. In fact, some people distinguish the top three layers of the OSI model as "transport layer users" and the bottom four layers as "transport layer providers."

There are five classes of transport protocol services:

Class	Title	Type
0	Simple	A
1	Basic error recovery	B
2	Multiplexing	A
3	Error recovery & multiplexing	B
4	Error detection & recovery	C

Class 0, known as Telex, is the simplest quality of service. It assumes that the network layer (below the transport layer) provides flow control. It releases its connection when the network layer releases its connection. Class 1 service was developed by the CCITT for its X.25 packet-switched network standard. It does provide expedited data transfers, but still relies on the network layer for flow control.

Class 2 represents an enhanced class 0. The basic assumption made here is still that the network is highly reliable. The quality of service offered includes the ability to multiplex multiple transport connections from a single network connection. Class 2 ensures that multiplexed packets of data arriving out of order can be reassembled properly.

Class 3 provides the services offered by both 1 and 2, as well as the ability to resynchronize so a connection can be reestablished if an error is detected. Finally, Class 4 assumes that the network layer service is inherently unreliable. It offers its own error detection and recovery procedures.

The network layer

The network layer is where network routing takes place. As such, it's key to understanding how gateways to IBM mainframe and other computer systems function. While upper-level OSI protocols request that a packet be transmitted from one computer system to another, the network layer concerns itself with the actual mechanics of the journey.

The network layer also forms the basis of the CCITT's X.25 standard for wide area networks. Later in this book you'll examine the actual structure of an X.25 packet including how the control information is packaged in several different fields.

The network layer performs a number of key services for the transport layer immediately above it in the OSI model. It notifies the transport layer when it detects unrecoverable errors and helps this layer maintain quality of service and avoid network congestion by stopping the transfer of packets whenever necessary.

Because physical connections can change from time to time when two networks are communicating, the network layer maintains virtual circuits and ensures that packets arriving out of sequence are reassembled correctly. In effect, it uses routing tables, which help it determine which path a particular packet should take. Often a message composed of multiple packets will take different pathways. The network layer provides essential "shipping" information for these packets, such as the total number of packets composing a message and the sequence number of each packet.

One very unfortunate complication in network communications is that different networks have different-sized address fields, different-sized packets, and even different time intervals during which they permit a packet to circulate before they "time out," the packet is considered lost, and a duplicate packet is requested. The network layer must provide enough control information within the packets to address these issues and ensure successful delivery and reassembly.

As mentioned earlier, there's a good deal of duplication between the functions of the transport layer and the network layer, particularly in flow control and error checking. The primary reason for this duplication is that there are two different types of connections, *connection-oriented* and *connectionless*, and they make far different assumptions regarding network reliability.

A connection-oriented network functions very much like our telephone system. Once a connection is established, communication continues, point to point, without the two parties feeling compelled to conclude each statement with their name, the name of the party they called, and that party's address. It's assumed that the communication is reliable and that the other party receives the message as it's sent.

Given the assumption of a reliable, connection-oriented network, the destination address is required only when the connection is established; individual packets don't have to carry this address in a separate field. The network layer assumes responsibility for error checking, as well as flow control, in a connection-oriented network. It also concerns itself with the actual sequencing of the packets.

A connectionless service, on the other hand, relies more on the transport layer for error checking and flow control. Destination addresses are required on each packet, and packet sequencing is not guaranteed. The basic assumption of this service is that speed is paramount and end users should have their own error-checking and flow-control software rather than rely on a standard built into the OSI model.

As is usually the case whenever committee members argue a complex issue, a compromise was reached that really didn't totally satisfy anyone. The compromise is that the capabilities for both connection-oriented and connectionless service are both built into the OSI model's network and transport layers. End users can select appropriate default values for these two layers' control fields and use the approach they prefer. The negative side to this compromise, however, is the significant amount of redundancy built into these two layers, which means a significant amount of bit overhead. When information using this OSI format is sent over long-distance lines, it translates into increased costs since the transmission takes longer.

The data link layer

The data link layer can be likened to the warehouse and receiving/shipping dock of a large manufacturing company. It must take the packets it receives from the network layer and prepare them for transmission (shipping) by placing them in the appropriately sized packages (frames). When information is flowing up the layers of the OSI model, the data link layer must be able to take the raw bits coming from the physical layer and make sense of them. It must establish where a transmission block starts and where it ends, as well as detect whether there are transmission errors. If it does detect errors, this layer is responsible for initiating action to recover from lost, garbled, and even duplicate data.

Several data links may exist simultaneously and function independently between computer systems. The data link layer is also responsible for ensuring that these transmissions don't overlap and that data doesn't become garbled. The data link layer initializes a link with its corresponding layer on a computer with which it will communicate. It has to ensure that both machines' clocks are synchronized and that they both use the same encoding and decoding schemes.

Since flow control and error checking are also the responsibility of the data link layer, it monitors the frames it receives and maintains statistical records. Upon the completion of a user's transfer of information, the data link layer assumes responsibility for determining that all data was received correctly before terminating the link.

Error checking in the data link layer. There are at least three different error-checking methods related to the automatic repeat request (ARQ) techniques used by the data link layer to perform this function, depending on the data link protocol it happens to be running. Stop and wait ARQ is a method in which a computer transmits a frame of information and then waits to receive an acknowledgment (ACK) control code indicating that the frame was received correctly. If an error is detected, a negative acknowledgment (NAK) is transmitted by the receiving station and the sending station responds by repeating its transmission.

The Go-back-N continuous ARQ approach enables a station to receive several frames (depending on the protocol used) before replying with either an ACK or a NAK pinpointing which frame contained an error. If a station sent seven continuous frames and an error was detected in frame number 4, then the sending station would reply to the NAK by retransmitting frames 4 through 7.

The selective-repeat continuous ARQ method provides an enhancement to go-back-N continuous ARQ. A receiving station maintains a buffer to store all frames received in a sequence and then replies that a particular frame, let's say number 4, contained an error. The receiving station keeps the other frames in a buffer and sends a NAK. The sending station retransmits only the frame containing an error (number 4 in this example). The receiving station then reassembles the frames into their proper sequence (1 through 7) and processes the information.

Key protocols for the data link layer. The data link layer contains a number of protocols that have been defined by the IEEE's 802 committee. In order to understand this key OSI model layer, you need to understand some of the work by this committee. These IEEE protocols are discussed in the next section, *IEEE and Network Standards*.

The physical layer

The physical layer of the OSI model is the least controversial since it contains international hardware standards that have become commonplace. In fact, one real issue concerning this layer is how the ISO will handle emerging new hardware standards. As transmission methods become faster and new interfaces develop with additional error-checking functions, will standards be added to the OSI model or will the physical layer remain static? At this point, the jury is still out and it's impossible to predict the ISO's response.

The physical layer spells out in great detail what kinds of pluggable connectors are acceptable. It lists, for example, 25-pin connectors for RS-232C interfaces, 34-pin connectors for CCITT V.35 wideband modems, and 15-pin connectors for public-data network interfaces found in CCITT recommendations X.20, X.21, X.22, etc. It also spells out what electrical characteristics are acceptable, including RS-232C, RS-449, RS-410, and CCITT V.35.

The physical layer is able to handle both the asynchronous (serial) transmission used by many PCs and some low-cost LANs, and the synchronous transmission used by some mainframe computers and minicomputers.

Since the ISO and the IEEE subcommittees have worked together so closely for the past few years, it isn't surprising that many LAN standards use the definitions provided at the physical layer by the OSI model. Various IEEE subcommittees have developed detailed descriptions regarding the actual physical equipment that transmits network information as electrical signals, including specifics on the types of cabling, plugs, and connectors to be used using the OSI model's physical layer as its basis for description.

The OSI model's physical layer defines such key network components as the type of baseband coaxial cable required to achieve 10-Mbps transmission speed. The layer incorporates an IEEE 802.3 definition for a thinner type of baseband coaxial cable for an approach known as cheapernet. The

IEEE 802.3 definition of a 10-Mbps twisted-pair baseband standard will also be added to this physical layer.

Under the physical layer's area of responsibility, you'll also find definitions for the type of fiber-optic and twisted-pair cabling required for a wide range of LANs. Some networks, such as IBM's Token-Ring Network, work with unshielded twisted-pair cable, while other types of networks might require shielded twisted-pair cable. The subcommittee has also defined the various types of coaxial cabling required for a variety of broadband local area networks.

The physical layer defined by the OSI model subcommittee must also specify the type of encoding scheme a computer uses to represent binary values for transmission over a communication channel. Ethernet, as well as many other local area networks, use Manchester encoding. With this approach, all network workstations are able to recognize that a negative voltage for the first half of the bit time followed by a positive voltage for the second half of the bit time represents the value 1, while a positive voltage followed by a transition to a negative voltage represents the value 0. There is a transition, therefore, from negative to positive or from positive to negative during every bit time.

You've seen that the physical layer assumes responsibility for the type of physical media, type of transmission, encoding method, and data rate associated with different types of local area networks. It's also responsible for establishing the physical connection between the two communications devices, generating the actual signal and then making sure that the two devices are synchronized. The timing of the two units' clocks must be the same so transmitted information can be decoded and understood.

Figure 2.3 summarizes the key protocols associated with the lower four OSI layers. These layers provide the connections that link heterogeneous LANs together. Notice particularly the redundancy built into the OSI model when it comes to connection-oriented and connectionless modes.

Transport layer	Transport service definition Connection-oriented transport protocol		
Network layer	Connectionless-mode network service		
Data link layer	Logical link control Unacknowledged connectionless service Connection-oriented service Acknowledged connection service		
Physical layer	CSMA/CD Baseband coax Broadband coax Unshielded twisted pair (1Mbps) 10BaseT (10 Mbps)	Token bus Broadband coax Carrierband coax	Token ring Shielded twisted pair Optical fiber

Figure 2.3 The LLC frame.

IEEE and Network Standards

The IEEE is an international professional society whose activities are organized around a series of boards, including a Standards Board. The Standards Board, accredited by the American National Standards Institute (ANSI), submits the standards it develops to ANSI for approval as national standards. Let's take a few moments to examine what some of these standards are and how they define the various tasks that networks perform.

The IEEE 802 committee

The IEEE established an 802 committee in 1980 to develop a set of standards for local area networks. Its goal was to develop standard interfaces that would make it cost-effective for manufacturers to develop products for new types of technology. The IEEE and ISO work closely together, so IEEE standards are incorporated into the OSI model. The 802 committee is currently divided into the following subcommittees:

802.1	Higher Layers and Management (HILI)
802.2	Logical Link Control
802.3	CSMA/CD Networks
802.4	Token Bus Networks
802.5	Token Ring Networks
802.6	Metropolitan Area Networks
802.7	Broadband Technical Advisory Group
802.8	Fiber Optic Technical Advisory Group
802.9	Integrated Data and Voice Networks

IEEE 802.1 and 802.2 standards

In this section we'll examine some of the issues related to the 802.1 and 802.2 subcommittees because their work is basic to understanding how local area networks function. Since the committees developing the OSI model and the IEEE committees have worked together closely, you'll see that these IEEE 802 standards use the same terminology as the OSI material you've just read. The IEEE 802 standards will provide a foundation for later discussions of how bridges and gateways function.

The IEEE 802.1 high-level interface for local area networks

The IEEE 802.1 committee is responsible for developing standards for bridges and has adopted a spanning tree standard, which is the method DEC has used for its Ethernet bridges. Another subcommittee (802.1D) is looking at the possibility of adding IBM's approach to bridging (source routing) as a standard. We'll look at these various approaches in a later chapter when we examine bridging.

While an army can travel on its stomach, an international standards organization travels on the seat of its pants, with committees breaking into a number of subcommittees. One 802.1 subcommittee (802.1A) is responsible for providing a network management architecture consistent with the OSI model. Still another subcommittee (802.1B) is developing network management protocols. We'll be looking at network management protocols in chapter 11, comparing and contrasting them with current network management approaches.

The IEEE 802.2 logical link control and medium-access control sublayers

The IEEE 802.2 subcommittee had to define a number of LAN functions. The layer above the physical layer, corresponding to the OSI model's data link layer, was to be responsible for such key functions as assembling data into a frame with address and error-detection fields, and, at the receiving end of a communications link, disassembling the frame, recognizing the address, and detecting any transmission errors. The subcommittee divided the second layer into two sublayers—logical link control (LLC) and medium-access control (MAC)—based on the specific functions performed. The LLC is independent of a specific access method and the MAC is protocol-specific; the result of this split makes it much easier for network designers to provide more flexible local area networks.

The IEEE 802.2 standard provides for both connectionless and connection-oriented services. Connectionless service is the norm in a local area network because of the high rate of speed and reliability. Type 1 LANs under this standard don't use the logical link control layer to provide for error checking, flow control, and error recovery. Instead, they use the appropriate protocols found under the transport layer of their network operating system software.

A type 2 connection-oriented service is also offered under the LLC standard. This service provides acknowledgments (ACKs) for error checking, as well as flow control and error recovery.

The logical link control sublayer

The primary purpose of the LLC sublayer is to enable network users to exchange data across the transmission channel controlled by the MAU. A connection-oriented LLC service contains primitives to control the establishment of a connection, transfer of data, and termination of a connection. A connectionless service, on the other hand, doesn't provide any error checking or acknowledgment that data arrived safely; it usually relies on transport layer services for this task.

An LLC frame, as illustrated in Figure 2.4, includes both a source and a destination address, as well as control information. The commands exchanged by the corresponding LLCs on the two computers that establish communications with each have a control field one byte in length. These eight bits can provide commands for connection, disconnection, acknowledgment, and rejection of frames. With a connection-oriented service, the LLCs send sequence numbers so the frames can be reassembled in the proper order. Each LLC maintains a send counter and a receive counter to keep track of frames sent and received and check for errors.

Figure 2.5 illustrates how LLC sublayers are involved in any communication between workstations under the IEEE 802 scheme. Say an application process on computer A wants to communicate with an application process running on computer B. When this information reaches the network layer on computer A, it's converted into an L_CONNECT.request, which is given to the LLC sublayer. The LLC sublayer passes this request on to computer B. Assuming that the connection is acceptable, the network layer on computer B issues an L_CONNECT.confirm, which ultimately reaches the network layer on computer A. Once communication is established, the two computers can exchange data using L-DATA_CONNECT request, indication, and confirm primitives. When they're finished communicating, the computers can use the appropriate request, indication, and confirm primitives associated with L_DISCONNECT.

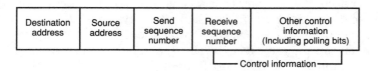

Figure 2.4 How the LLC sublayers communicate.

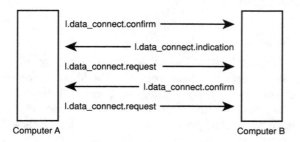

Figure 2.5 Protocols associated with the four lower layers of the OSI model.

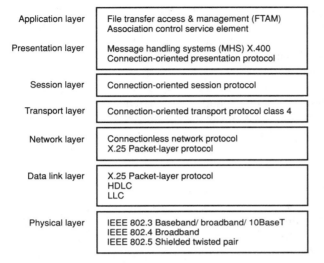

Application layer	File transfer access & management (FTAM) Association control service element
Presentation layer	Message handling systems (MHS) X.400 Connection-oriented presentation protocol
Session layer	Connection-oriented session protocol
Transport layer	Connection-oriented transport protocol class 4
Network layer	Connectionless network protocol X.25 Packet-layer protocol
Data link layer	X.25 Packet-layer protocol HDLC LLC
Physical layer	IEEE 802.3 Baseband/ broadband/ 10BaseT IEEE 802.4 Broadband IEEE 802.5 Shielded twisted pair

Figure 2.6 GOSIP protocols.

The medium-access control (MAC) sublayer

The medium-access control (MAC) sublayer defines how different stations can access the transmission medium. The LLC frames are nonhardware-specific. The MAC must ensure that it provides information in the appropriate form for a bus, token-bus, or token-ring network. The OSI model's data link layer functions of addressing (both source and destination), error detection, and framing also take place at this level.

The Government OSI Profile (GOSIP)

August 15, 1990 is a significant date in computer history, especially if your favorite subject is protocol. On this date, the U.S. government made the government OSI profile (GOSIP) a mandatory requirement for networking procurements. GOSIP is a subset of the OSI model protocols, as reflected in Figure 2.6.

GOSIP will be implemented in at least three different versions. The first version will permit government computers to use the application layer's electronic mail service (message handling services X.400) and its file management service (file transfer, access, and management, or FTAM) in addition to the IEEE logical link control protocols for the popular bus, token-bus, and token-ring networks. Notice that GOSIP includes the X.25 packet layer protocol, which means that both wide area networks and local area networks will be able to communicate together. Notice also that both connection-oriented (transport layer) and connectionless (network layer) services are provided.

The second version of GOSIP is expected to add a protocol known as office document architecture (ODA) above the application layer. This protocol will enable software to manipulate the paragraphs, sections, chapters, and pictures that comprise a document and treat each of these document components as a distinct object.

GOSIP's third version is expected to add X.500 global directory service, the importance of which I'll explain in chapter 11 when discussing electronic mail on a network. Other protocols on the drawing board for GOSIP include FDDI, a topology discussed in chapter 1, and electronic data interchange (EDI), a way to exchange documents and their contents electronically.

TCP/IP: today's protocol standard

Almost every computer industry expert predicts that the OSI suite of protocols will become the standard for internetwork connectivity in the near future, but what about today? The current de facto standard for internet connectivity is transport control protocol/internet protocol, which is better known as TCP/IP.

TCP/IP was developed almost two decades ago at the request of the Department of Defense's Advanced Research Projects Agency (ARPA), and is incorporated in the government network known as the Defense Data Network (DDN), which includes ARPANET and MILNET. It also serves the Internet, a network of networks linking over 800 networks and 120,000 host computers. In this section, I'll explain TCP/IP's various protocols and its advantages and disadvantages, and then describe the issues associated with a company migrating from TCP/IP to compliance with OSI standards.

TCP/IP is a set of layered protocols just like the OSI model. As Figure 2.7 illustrates, TCP and IP correspond roughly to the OSI's transport and network layers. Since TCP/IP is concerned primarily with transport and network services, it isn't limited to specific hardware platforms. It can run over X.25, the data link protocol comprising the bottom three layers of the OSI model, commonly used for packet-switching networks—which you'll examine in chapter 7.

TCP/IP began to grow in the private sector in 1982 when the UNIX 4.2 BSD operating system incorporated TCP/IP into its kernel (the very heart of the operating system). This married TCP/IP and UNIX, which made TCP/IP's protocols free and readily available to all UNIX networks. A second major event in 1982 for TCP/IP was the incorporation of the address resolution protocol (ARP). ARP is very significant because it maps Ethernet to TCP/IP internet addresses. Imagine the significance of these two events for the business community. A company running UNIX on its minicomputer networks and Ethernet on its LANs had a viable way of connecting the two worlds seamlessly.

More recently, Sun Systems has taken advantage of TCP/IP. It uses its own network file system (NFS) protocol on its Sun workstations, which

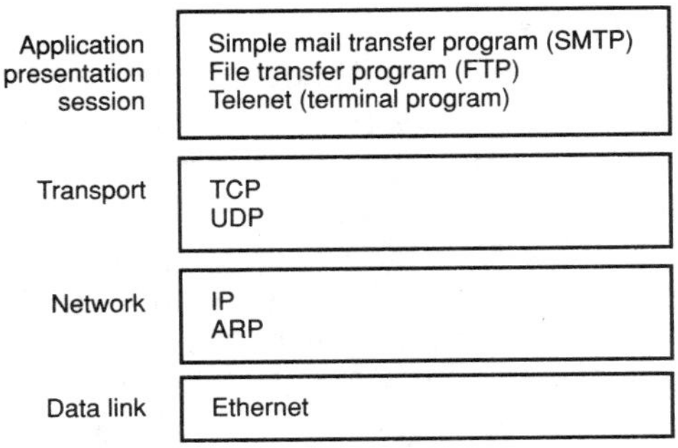

Application presentation session	Simple mail transfer program (SMTP) File transfer program (FTP) Telenet (terminal program)
Transport	TCP UDP
Network	IP ARP
Data link	Ethernet

Figure 2.7 TCP/IP protocols and services, and the OSI model.

dominate the workstation market, and it uses TCP/IP in conjunction with NFS to provide internet communications. In fact, it's hard to find hardware or operating systems that don't have a TCP/IP implementation. Versions run on IBM Token-Ring LANs as well as Ethernet LANs, Macintosh computers, and DEC's VMS operating system.

Network services under TCP/IP

You'll notice in Figure 2.7 that TCP/IP also provides network services through its Telnet, FTP, and SMTP applications. Telnet offers virtual terminal service, which means it provides a network-standard terminal to which other terminals can be mapped. In effect, hosts can exchange information using a generic terminal mode and then each host is responsible for mapping the generic terminal to the terminals at its location. The advantage of such an approach is that a network terminal in Washington really doesn't need to know or care about what kind of terminals are running on the computer network in Los Angeles.

While the majority of terminals using Telnet select the full-duplex ASCII character code set, an EBCDIC mode is available for communicating with IBM mainframes and their terminals.

File transfer protocol (FTP) is the TCP/IP service for file transfers. It enables two heterogeneous networks to exchange files in binary, ASCII, or EBCDIC format and permits unattended transfers, the most common mode of file transfer, from remote sites. To effect a file transfer, a user would provide FTP commands indicating the type (ASCII, EBCDIC, or image), structure (important if files have a page structure), and mode of files (sending data as a stream or sequence of bytes is one option, while block and compression are the two other choices).

The simple mail-transfer protocol (SMTP) is the TCP/IP service for electronic mail. It's flexible enough to handle distribution lists as well as mail forwarding for computers not connected to the internet using TCP/IP.

The OSI suite of protocols offers more advanced versions of these services, which I'll explain in chapter 11 when we look at network management.

There are several reasons these TCP/IP services are so appealing, though. They're virtually free. Equally important, they're found on tens of thousands of network nodes and work efficiently if not spectacularly. They still represent the easiest "quick and dirty" way to provide file transfer, terminal emulation, and electronic mail services for linked heterogeneous networks.

The best way to see how TCP/IP works is to follow data being sent from one computer network to another network using this suite of protocols. Transmission control protocol (TCP) corresponds roughly to the OSI's transport layer, but also contains some session layer functions. It's responsible for establishing a session between two user processes on the network. It's also responsible for error recovery, and is primarily responsible for encapsulating information into a datagram structure, transmitting the datagrams, and keeping track of their progress. It handles the retransmission of lost datagrams and ensures reliability. Finally, on the destination computer, TCP extracts the message from the datagram structure and forwards it on to the destination application program.

TCP adds a header containing its control information to any data coming from an application program to create a datagram. The internet protocol (IP) adds its own header containing its own instructions to the datagram. Finally, the local network adds its own local network control information in the form of its own header to the datagram. Figure 2.8 illustrates how a datagram includes three distinctly different headers, each of which contains different control information.

The transmission control protocol (TCP) takes data it receives from an application program and encapsulates it into a datagram that includes its own header, which contains its control information. Figure 2.9 illustrates this datagram structure. The source and destination port fields represent the station calling the TCP and the station being called. The sequence number keeps track of the order of the datagrams to be received so the data is received in the correct order. The acknowledgment number field indicates the next byte sequence that's expected.

The data offset field tells you the number of 32-bit words in the TCP header. You don't want to confuse the information in the header with the data field. The flag section gets more technical than you need at this point, but among the functions represented are an indication of how urgent this particular datagram is, as well as notice of a session's setting up or closing down.

The window field indicates the number of octets of information the sender is willing to accept or receive. The checksum field is a 16-bit number that in-

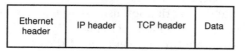

Figure 2.8 A TCP/IP datagram with its three headers.

Source port		Destination port
Sequence number		
Acknowledgment number		
Data offset	Flags	Window
Checksum		Urgent pointer
Options		Padding

Figure 2.9 A TCP datagram structure.

dicates whether the datagram is received intact. The urgent pointer contains the number of octets from the beginning of the TCP segment to the first octet following any urgent data. The options field can contain a variety of information, including the maximum TCP datagram size. Finally, the TCP header includes some zeros to pad the datagram to the next multiple of 32 bits.

User datagram protocol (UDP)

Sometimes TCP is overkill. There are occasions when there's no need for the reliable transport services provided by this protocol. The TCP/IP suite of protocols also includes user datagram protocol (UDP), a protocol designed for applications that don't need the firepower of several datagrams sequenced together. You might want to think of TCP as Federal Express, with its elaborate computerized tracking of packages and its promise of reliability. UDP is more like regular mail delivered by the U.S. Postal Service. It's inexpensive in the sense that it doesn't require a lot of overhead (extra bits for all TCP's fields). It doesn't split large packages into smaller packages, and it doesn't keep track of what arrives safely and what needs to be resent.

The UDP header is shorter than TCP's, but it still includes a source and destination address. It has a checksum field to check for errors, and that's about it. This is used by other protocols that need to look up something short, such as a name from a name table.

Internet protocol (IP)

Internet protocol serves as the router for the datagrams. This protocol concerns itself with issues such as fragmentation of datagrams and internet addressing. Since different networks have packets of varying size, datagrams might have to be fragmented so they don't exceed the maximum packet size. IP provides the control information necessary to reassemble these fragmented datagrams.

A second major task of internet protocol is internet and global addressing. Three different types of addressing schemes are available, depending on the size of the network to which the datagram will be routed. The best way to understand the IP is to examine its datagram structure.

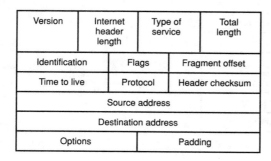

Figure 2.10   An IP header.

Figure 2.10 illustrates an IP header. The version field indicates which version of IP is being used and, therefore, the header's format. The header length field is its length in 32-bit words. The type of service field indicates the quality of service desired. The parameters for this field provide information on the datagram's precedence (importance), intervals (will there be a steady stream of these datagrams at regular intervals in the future?), level of reliability (are mistakes crucial?), the importance of delivery speed of this datagram, and an indication of the relative importance of speed versus reliability in the event of a conflict between the two.

The quality of service field is followed by the total length field, which is similar to the same field found in the TCP header. It represents the total length of the datagram, including its IP header. The identification field contains a unique number for identifying this particular datagram, while the flags and fragment offset fields include information on whether a datagram is fragmented, if more fragments are still coming, and the number of octets from the beginning of the datagram for a specific fragment in 64-bit units.

The time to live (TTL) field records how long the datagram is permitted to exist in the network. This number is decremented each time the datagram is processed along its journey through the network, and discarded when the number reaches zero. The maximum number permitted is 255 seconds; the point of having a TTL field is to preserve the resources of the network and avoid time delays due to a large number of undelivered datagrams.

The protocol field following the TTL field indicates which protocol follows the IP. If there's a TCP header, its number would be found in this field. The functions of the header checksum, source address, and destination address are self-explanatory, and they serve the same purposes as their counterparts in the TCP header, which I've already discussed. Finally, the options field can contain information on tasks such as routing specifications.

TCP/IP Versus OSI: The Aging Champion and the Challenger

You've looked at the current leader in internet connectivity (TCP/IP) as well as the OSI model that most experts predict will become the future leader. As a systems integrator, what do you do if your company needs to link disparate networks now? What if you have a UNIX-based minicomputer network in Los Angeles, a VAX Ethernet network in Chicago, and a NetWare network running in Boston? Right now you could run TCP/IP protocols on all these networks, which means you could have internet corporate electronic mail (SMTP) as well as file transfers between sites (FTP).

There are limitations to TCP/IP. Its mail program is very basic and lacks the bells and whistles of OSI's X.400 mail service. Its file transfer program is also much more limited than OSI's FTAM.

More significantly, it's ironic that a task force of the American National Standards Institute (ANSI), the X3S3.3 Committee, is moving this venerable suite of protocols toward official standardhood. It's almost like honoring an actor with an academy award even though the actor's best work was several years ago. TCP/IP will not grow and evolve. As pointed out in the GOSIP section, emerging OSI standards will provide support for electronic data interchange as well as sophisticated network management functions. A company that knows it will need these features in the future will need OSI protocols.

Strategies for Moving Toward OSI

Okay. Your company bit the bullet and implemented TCP/IP a couple of years ago because you couldn't wait for OSI to develop fully and software publishers to start releasing OSI software. Your job is now on the line because your boss keeps reading about the OSI model and how antiquated TCP/IP is. How can you show that you've considered the need to ultimately move to OSI and have come up with viable alternatives to simply scrapping what you have and starting over?

The gateway approach

One solution to linking TCP/IP and OSI is to use a gateway. A gateway is much too sluggish for real-time network applications, but it can be adequate for mail service and other store-and-forward applications. Gateways aren't good for large file transfers or other activities that result in traffic jams and backups. Many companies might opt for this approach if their primary network communications for the foreseeable future will be over TCP/IP, with just occasional communications needed with OSI networks.

Several companies have been working on creating TCP/IP-to-OSI gateways, including Retix, Wollongong, and Touch. You can currently purchase FTAM-to-FTP and X.400-to-SMTP application-layer gateways.

The dual protocol stacks approach

A similar approach to bridge the TCP/IP-to-OSI gap is the dual protocol stack. With this approach, each machine on a network has an OSI protocol

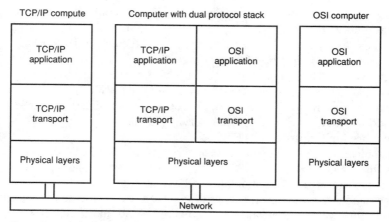

Figure 2.11 A dual protocol stack.

stack and a TCP/IP protocol stack. In effect, you have two completely different networks running over the same cabling. This approach offered by several software companies, including Wollongong and Touch, is illustrated in Figure 2.11.

Companies have chosen this approach because it's "quick and dirty" to implement and uses fairly simple technology. Drawbacks include the need for extra memory and processing power.

A third way to bridge the gap between TCP/IP and OSI is to use a transport-service bridge. This type of bridge, already offered by the Wollongong Group and Retix, allows you to run OSI applications over TCP/IP networks by acting like a router. It routes the OSI protocol data units by packaging them so they emulate TCP/IP. The RFC1006 standard is part of the OSI model's suite of protocols and is defined under the International Standards Organization Development Environment (ISODE).

A major advantage of using a transport-service bridge is that a dual protocol stack is required for only the first three layers. Networks can take advantage of their existing TCP/IP network links while adding OSI's upper layers and their functionality. The price you pay for this approach, however, is that it's less reliable than the transport layer. Since you're routing data by repackaging it to emulate a different protocol and not actually translating from one protocol to another, you wind up with a TCP checksum at one end and an OSI TP4 checksum at the other end. The two checksums are incompatible, so it's impossible to use this normal error checking tool.

Xerox Network Systems (XNS) Suite of Protocols

What is XNS? A decade ago, when micros were just starting to emerge as serious business machines, two roads diverged in the network woods. The government and universities, both of which were havens for mainframes and minicomputers, embraced TCP/IP's suite of protocols. The LAN industry leaders at that time (3Com, Novell, and Ungermann-Bass) all embraced a suite of protocols called Xerox Network Systems (XNS). More recently, Banyan Systems has used XNS protocols as part of its VINES network operating system.

LAN manufacturers opted for XNS rather than TCP/IP for a number of reasons. The major reason was that XNS was designed specifically for LANs. It cut overhead (compared to TCP/IP) by reducing the amount of error checking required by upper-level protocols. The greater reliability of LANs compared to internet communications meant that they could get away with this tradeoff.

Xerox never did develop conformance tests for its XNS. Since the LAN industry developed during a time when every manufacturer was striving for the best proprietary hardware and software system, the lack of a standard XNS didn't seem to be much of a problem at that time. After all, Novell, 3Com, and Ungermann-Bass were selling essentially closed systems. No one ever expected to link together LANs running different protocol suites.

It's worth spending a few moments with XNS, though, because several of its protocols are still used (albeit significantly modified in the case of NetWare). XNS uses a layered approach for grouping its protocols according to functions that became the basis of the OSI model. Figure 2.12 illustrates how XNS compares to OSI.

Each of XNS's layers provides services to the next higher level. Its top layer (4) corresponds to the OSI model's application layer. It contains application protocols implemented for specific platforms. The clearinghouse service, for example, is an application-level directory service that links the names and aliases of users to the corresponding network and data link addresses. An authentication service in this level allows specific users to access specific resources while barring others. Unfortunately, given the different implementations of XNS, these protocols lack the universality of OSI's X.400 and X.500 protocols (see chapter 11 for a more detailed exam-

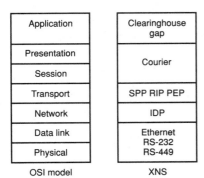

Figure 2.12 XNS and the OSI model.

ination of these emerging standards). The courier layer provides services similar to OSI's session and presentation layers.

Since XNS was developed specifically for LANs, its major emphasis is on the network and transport layers. At the network layer, XNS uses an internet datagram protocol (IDP), which specifies a maximum packet size of 576 bytes. Various proprietary XNS versions use IDP, which specifies a different-sized packet. This illustrates the difficulties in linking these network "cousins" together. For its NetWare product, for example, Novell has taken IDP and modified it to create the internet packet exchange (IPX) protocol, which is no longer compatible with IDP.

The very heart of XNS is its suite of transport-layer protocols for routing packets across the network. Since routing packets is the primary job of an LAN, let's take a look at how XNS (and 3Com's LANs, which use it) handles routing. I'll also touch on some ways that Novell has taken XNS and modified its protocols to enhance its own performance. You'll return to this topic in chapter 4 when you look at several other types of routing approaches, as well as devices that function specifically as routers to link together networks using different protocols.

XNS in action

XNS's key transport protocols include routing information protocol (RIP), sequenced packet protocol (SPP), and packet exchange protocol (PEP). Let's examine how XNS routes a datagram from one network to another. The closest analogy I can give you is the way a tourist finds a specific address in Tokyo. Because of the haphazard addressing scheme found in Tokyo, where buildings are given addresses based on when they were built rather than their physical location on a block, the normal way to locate a specific building is to go from police station to police station and inquire. Each station handles a specific neighborhood. The officer will provide you with specific directions to a police station closer to your target. Finally you arrive at the police box for the appropriate neighborhood, where you receive directions to the actual building.

Now let's note the similarities in the way XNS and its RIP protocol handle the transport of a datagram from one network to another. The source node where the packet originates checks a table to determine the internet address of the destination node. It might also broadcast a request across the network, "Does anyone know where Bill Taylor lives?"

If the destination address is on the same network, the process is very simple. IDP adds its header to the packet containing the data and address, and passes the information down to layer 0 (equivalent to OSI's data link layer) for delivery.

Assume, though, that you're sending a packet from one network to another. IDP takes the packet and places its network router address on it. This router workstation keeps detailed address tables using RIP protocol. The router looks at these tables and determines the best route for the packet. It then places the address of the next router on that route (removing its own address in the process) and sends the packet along its way.

This process continues as many hops as necessary until the router on the network containing the destination node receives the packet. It forwards this packet directly to the appropriate destination workstation.

XNS routers "listen" to network traffic and update their routing tables every 30 seconds. If a router can't get information to a specific LAN, all other routers try. If they ultimately fail, there's general agreement that the LAN isn't capable of responding. RIP stipulates that each router broadcast its address information to all LANs, including the LANs from which it received the information.

Vendors such as Novell have modified their versions of RIP to make the protocol more efficient. NetWare broadcasts its routing information every 60 rather than 30 seconds, which reduces some traffic. It has also increased the RIP packet size to include room for information to keep packets from being discarded when they fail to complete their journey in the time RIP normally allocates.

What happens under XNS if packets arrive out of sequence? The sequenced packet protocol (SPP) ensures reliability above the simple datagram delivery of IDP by synchronizing sending and delivery using a full-duplex communication method. Since packets are numbered, lost or damaged packets can be requested and then retransmitted. NetWare has taken SPP and modified it to create its sequential packet exchange (SPX) protocol.

Summary

Protocols are rules that enable devices to communicate with each other. The open systems interconnect model (OSI) is a model of layered protocols developed by the International Standards Organization to facilitate communications among heterogeneous networks. Transport control protocol/internet protocol (TCP/IP) is a suite of protocols developed for the U.S. Defense Department's network and is the foundation of our nation's internet. Industry experts believe that networks will eventually move from TCP/IP to OSI protocols. The most viable way to do so at present is to use a dual protocol stack.

Xerox Network Systems (XNS) is a suite of protocols that's the basis of many network operating systems, including NetWare. Its semiproprietary nature has hindered its development. The network operating system market has been dominated by Novell's NetWare, but Microsoft's NT Advanced Server is gaining momentum.

3

Local Bridges

In this chapter, you'll examine:

- How bridges filter and forward frames
- Spanning tree bridges
- Source routing bridges
- How spanning tree and source routing bridges communicate
- How the 802.1D standard provides seamless linking of spanning tree and source routing bridges

Some industry surveys indicate that as many as 75% of Fortune 1000 companies have token-ring and Ethernet LANs they want to link together. Bridges have become a key component of enterprise networks; they provide increased connectivity, security, and efficiency. In this chapter you'll see some of the more common types of local bridges and observe how their methods of operation differ. You'll see why it's a complex process to bridge Ethernet and token-ring networks. Finally, you'll learn how the IEEE 802.1D standard provides transparent bridging of these two very different networks.

The Internet or Enterprise Network

Anyone who has followed the local area network industry over the past five years knows that large companies no longer concern themselves with linking two small LANs together. Today, companies want to know how to link together several LANs that might or might not be located at the same physical site. Each individual network in such a system is known as a *subnetwork*, since it's part of a larger internet. Be sure not to confuse the term *internet* with the Internet, the national network that links Defense Department, university research, millions of corporate computers, and countless personal computers.

What Is a Local LAN Bridge?

A local LAN bridge consists of the hardware and software required to link together two different LANs or subnetworks located at the same site into one internet. I'll describe remote bridges in conjunction with wide area networks later in this book. The simplest type of bridge examines a packet's 48-bit destination address field and compares this address with a table that lists the addresses of workstations on its network. If the address doesn't match any of its workstations, it forwards the packet across the bridge to the next network. These simple bridges keep forwarding packets, hop by hop, until they reach a network containing a computer with the desired destination address. This process of examining address tables and forwarding packets is called *transparent bridging*; it's a technique used by all Ethernet bridges and some token-ring bridges. Figure 3.1 illustrates how this type of bridge operates.

Some bridges create their own network address tables. These bridges examine the source and destination addresses of every packet transmitted onto the LANs to which they're connected; they then build their own address tables, which list workstation source addresses of packets on their network that have this network's corresponding number. These bridges then try to match the destination addresses of packets with one of these source addresses. When a bridge matches an address, it *filters* the packet, sending it on its way along the network, where the destination workstation will recognize its own address and copy the packet to RAM. If there's no match, then the packet is *forwarded*, or permitted to travel across the bridge to the next network. Broadcast and multicast packets are always forwarded because their destination address fields are never used as source addresses.

Bridges don't understand or concern themselves with higher-level protocols. They function at the media access (MAC) sublayer of the OSI model's data link layer, far removed from upper-level protocols such as XNS or TCP/IP. As long as networks on both sides adhere to IEEE 802.2 logical link control (LLC) standards, a bridge can span them regardless of differences in their media or network access method. As you'll discover when reading this chapter, this means that it's possible for corporations to bridge their Ethernet and token-ring networks, as well as their 802.3 LANs, by using a 100BaseX Ethernet with data-grade twisted-pair wire, a 10BaseT Ethernet with unshielded twisted-pair wire, or a thin coaxial-cabled "cheapernet" network.

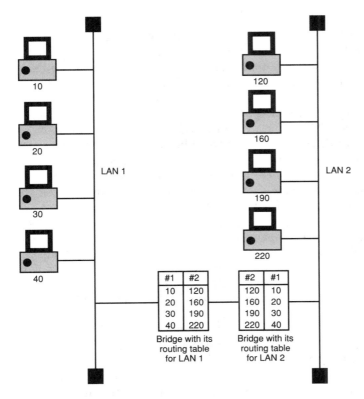

Figure 3.1 A simple bridge using transparent bridging.

Why Use Bridges?

There are a number of network design reasons for using bridges, including increased efficiency, security, and distance. Efficiency is usually the most often cited reasons for installing bridges on a network. Because bridges are capable of filtering packets according to programmable criteria, a network manager can use a bridge to reduce traffic congestion and improve speed by dividing up a large network and then bridging the resulting subnets. The two smaller networks would run faster because they had less traffic.

Since larger Ethernet networks are slowed down by collisions, it makes sense to create smaller Ethernet subnetworks and use a bridge to provide such services as e-mail. Ethernet has a maximum length limitation of 2.5 Km, and also restricts the number of network segments to three in order to avoid exceeding the 9.6-microsecond propagation delay. Network managers and systems integrators can overcome both Ethernet limitations by using bridges.

A 4-Mbps token-ring network limits the number of network workstations to 72 with unshielded twisted-pair cable and 270 workstations with IBM's own shielded cabling. Network managers can overcome these limitations by using smaller subnetworks and then bridging them. The smaller subnetworks operate more efficiently and are easier to manage and maintain.

Another reason why bridges make a network more efficient is that a network designer can use different topologies and media wherever appropriate and then link these different networks via bridges. Offices within a department might be linked via twisted-pair wire. A bridge could connect this network to the corporate fiber-optic backbone. Since twisted-pair wire is so much less expensive than fiber-optic cabling, the network design would save money and increase efficiency by using the high-bandwidth medium on a backbone where there's the most traffic.

Bridges can link two similar networks that have different transmission speeds. As an example, a company might be perfectly happy with an 802.3 1-Mbps StarLAN network using unshielded twisted-pair cable for one department, but require a 10base5 802.3 network using thick coaxial cabling and transmitting at 10 Mbps for its manufacturing plant. A bridge buffers packets, so it's no problem for it to span packets from LANs with different transmission speeds.

Since the 802 committee developed a common logical link control layer for its various network topologies, it's possible, for example, to link two token-ring networks separated by an Ethernet LAN. The Ethernet LAN can forward the packets just as a mailman can deliver a letter written in a different language, as long as the envelope (packet) follows the rules and regulations established by the standard.

While added efficiency is a major reason for using bridges, they can also increase security. They can be programmed to forward only those packets that contain certain source or destination addresses, so only certain workstations can send information to or receive information from another subnetwork. The accounting subnetwork, for example, can have a bridge that permits only certain workstations outside the network to receive information. In addition to providing a security barrier that filters out unwarranted access, bridges also add a measure of system fault tolerance. If a single file server on a large network fails, the entire network fails. If, however, internal bridges are used so that two file servers back up each other continuously, then traffic is reduced while providing an additional measure of security.

Finally, bridges can also increase the distance a network can span. Since a bridge rebroadcasts a packet to the workstations on the receiving side, it functions like a repeater to increase the distance a packet can travel without its signal attenuating. Often bridges are *cascaded* to connect LANs sequentially, as shown in Figure 3.2.

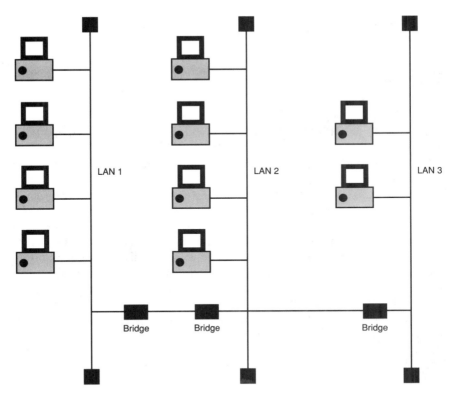

Figure 3.2 Cascaded bridges.

Intelligent Bridges

All bridges have the ability to update their routing tables; that's how they keep track of which workstations have been added to the network. Intelligent bridges differ from the simple bridges I've been describing by offering additional capabilities.

Intelligent bridges can be programmed to filter packets based on certain criteria. I referred to an application appropriate for this type of bridge when describing how a bridge can enhance network security by restricting traffic to and from specified subnetworks. An Ethernet segment for the accounting department might be connected to the rest of the network via a programmable bridge that permits only workstations on the corporate subnetwork to access it.

Intelligent bridges might offer source explicit forwarding (SEF). This feature enables network supervisors to assign internetwork access privileges by labeling specified addresses in a routing table as either accessible or inaccessible to specific users and groups.

Spanning Tree Bridges

The spanning tree algorithm (STA) was developed by DEC and Vitalink and later adopted as a standard by the IEEE 802.1 committee. Spanning tree is an approach toward bridging multiple networks where more than one loop might exist. Figure 3.3 illustrates how there can be multiple paths connecting one network with another. Data can be transmitted between LAN 1 and LAN 2 a number of different ways.

Under STA, each bridge has an identifier that consists of a priority field and a globally administered station address. Bridges negotiate with each other to determine the route data should take. A root bridge is selected during this negotiation process on the basis of having the highest priority value. If two bridges have the same priority value, the one with the higher station address is selected as the root bridge. While this process proceeds automatically, the network manager can "fix" the results by giving a particular bridge a higher priority value.

After the root bridge has been selected, each bridge determines which of its ports point in the direction of the root bridge and designates it as the root port. If more than one bridge is attached to the same LAN, a single bridge is selected based on the one that offers the least "cost," based on criteria established by the network manager. Costs could include elements such as line speed and buffer capacity. If all costs are set as equal, STA will produce a tree-like structure with bridge ports selected that result in the least number of hops from bridge to bridge for a packet to be transmitted from one LAN to another.

Now that all these negotiations have taken place, each bridge sets its root port in a forwarding state to move data toward the root bridge. It also sets its port pointing away from the root bridge in a forwarding state. Other bridge ports are blocked so that packets cannot travel through them. As Figure 3.4 illustrates, STA ensures that only one bridge port in each direction on a LAN is operating and that only the most efficient path is available.

What happens under STA if a bridge port goes out of service? A port that has been blocked is placed in a learning mode so it can examine packets flowing on the network and update its database of network station addresses. The port then changes to a forwarding mode and forwards a notification of its change to the root bridge. The root bridge notifies all network bridges to update their databases to include this new bridge port.

A bridge's address database includes information on the direction to forward data for a particular workstation, as well as a timer. If the timer for a particular workstation expires, then the information regarding which direction to forward data for this workstation might no longer be valid. The

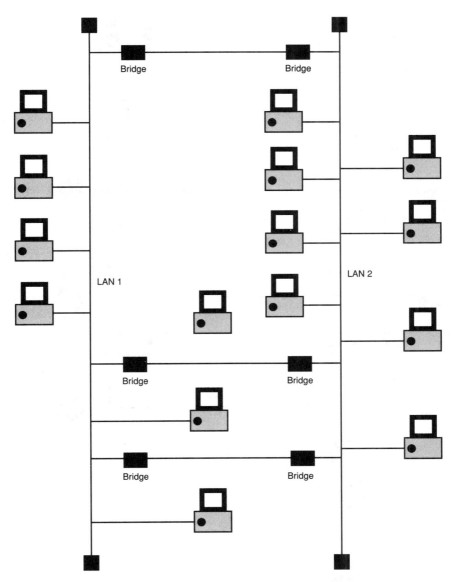

Figure 3.3 Bridges offering multiple paths between two LANs.

bridge monitors the source addresses on packets it receives and updates its address table. If a new network topology is introduced, it might require changing the direction to forward data information in a bridge's address database.

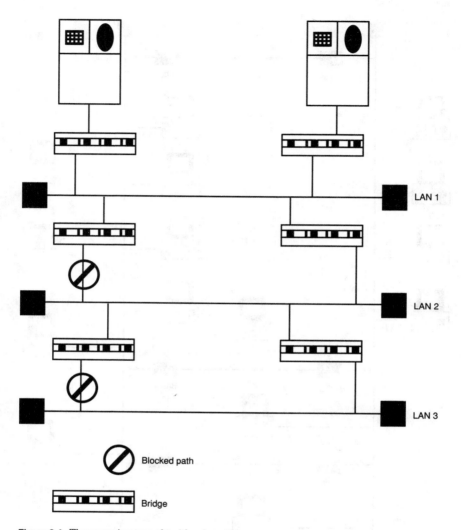

Figure 3.4 The spanning tree algorithm in action.

Source Routing Bridges

In 1985 IBM introduced source routing with its version of Token-Ring topology. In fact, IBM's PC LAN program and its OS/2 operating system are both designed to work only with source routing bridges. Source routing bridges actually perform routing duties at the network layer of the OSI model. Each device on a LAN that uses source routing must have a unique six-byte address. The address field's first high-order bit, called the I/G bit, indicates whether there's an individual or group address. The second high-order bit, the U/L bit, indicates whether the address is universally (IEEE addresses assigned to manufacturers) or locally administered.

Source routing takes the I/G bit in the source address only and uses it as the routing information (RI) indicator bit. When this bit is set to 1, it indicates the presence of additional routing information in the frame header. This additional routing information (up to 18 bytes in length) specifies the frame's complete path from source workstation to destination workstation. Figure 3.5 illustrates the location of this crucial routing information within a Token-Ring frame.

Each LAN ring is assigned a unique number just as an office routing slip might indicate the order in which a memo should be circulated before it reaches the file clerk for filing. If two bridges on the same LAN are parallel and can lead to the same destination, source routing will arbitrarily assign each of them a different number to keep the routing directions from becoming confusing. Figure 3.6 illustrates what this routing information would look like if it were translated into English.

The length of this information field limits routing to eight ring numbers, which means a maximum of seven bridges; source routing bridges refer to this limitation as a *seven-hop limit*.

How a networked PC gathers source routing information

A networked PC gathers source routing information by transmitting an all-routes broadcast frame to all rings connected on an internet. This frame contains control information and a blank buffer that can be filled in by other workstations. Bridges fill in the numbers for the two rings they connect and their own bridge number. The destination workstation receives this broadcast frame and returns it to the source station, which now has a road map of the route that the frame took.

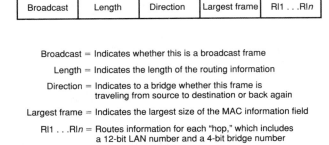

Figure 3.5 Source routing information within a token-ring network frame.

Control (2 bytes)	Ring 2	Bridge 1	Ring 7	Bridge 3	Ring 9	Bridge 0

Figure 3.6   A closer look at a source routing information field.

It's possible to use a spanning tree topology with source routing bridges. A special single-route broadcast frame is circulated once. It ensures that only certain bridges in the network are configured to pass single route frames. Since there are no loops permitted in a spanning tree topology, there can be only a single path between any two rings. Bridges won't pass a frame onto a ring if it has already circulated on that ring. A workstation uses the information it receives from its two types of broadcast frames to determine the optimal route containing the least number of hops for the frame it transmits.

The source routing bridge in operation

A source routing bridge examines every frame on each of the Token-Ring Networks it links. If it sees the RI bit set to one, it will examine the routing information field to see if the two ring numbers match the two rings it connects. Assuming the routing information matches, the bridge will forward the frame across the bridge. Frames that don't have matching ring numbers are filtered.

Current Ways to Link Ethernet LANs and Token-Ring LANs

An enterprise network by its very definition links together all computing resources in a company, including Ethernet and token-ring LANs. The accounting department might use Ethernet to connect PCs running productivity software such as WordPerfect and Lotus 1-2-3 to their VAX computer, which runs DEC accounting software. Other areas might use a token-ring network. What's a network manager or systems integrator to do in such a situation? You've already seen the differences in frame structure between the two networks and the significant differences between spanning tree and source routing approaches to routing frames.

It's absolutely crucial to remember that there's a significant difference between connectivity and interoperability. *Connectivity* refers to being able to link together the two networks and transmit data, while *interoperability* refers to the ability of each network to use the data transmitted to it.

Sometimes connectivity is all that's needed. Let's say you have several Ethernet networks in an enterprise network, along with a 16-Mbps token-ring network that serves primarily as a backbone, a gigantic switching station. While the 802.3 and 802.5 frames differ, they do have a common MAC layer. The token-ring network can forward Ethernet frames through its ring and onto a bridge connected to another Ethernet network. The token-ring network cannot "open" the frame and understand the data contained within it, but it can understand the source and destination address fields. What a token-ring-to-Ethernet bridge does is support source routing on the token-ring side and transparent bridging on the Ethernet side.

There are bridges available today that can perform the changes in the frame required to convert an Ethernet frame to a token-ring frame. With such bridges, workstations on the token-ring side view the token-ring bridge as just another bridge. Workstations on the Ethernet side, however, view the bridge as just another Ethernet workstation. Frames generated from the token-ring side and addressed to an Ethernet workstation are sent to the bridge, where they're stripped of the logical link control (LLC) protocol. They're then converted into Ethernet frames and transmitted over to the Ethernet network.

Frames sent from an Ethernet workstation to a token-ring workstation must go through an additional step. The bridge must search its own address database to learn the additional routing information required for source routing over token-ring networks.

The CrossComm token-ring-to-Ethernet bridge family was one of the first bridges to perform this crucial task. It supports higher protocols including NetWare, TCP/IP, and 802.3 LLC. As far as media, it supports thick and thin coaxial cabling, twisted-pair Ethernet and StarLAN, and fiber-optic Ethernet and token ring. The bridge is designed to detect Ethernet packets that don't have a source routing information field and insert this field so they can travel on the token-ring side of the bridge. The actual protocol conversion that takes place is handled by CrossComm's proprietary dynamic conversion mode technology.

IBM's 8209 LAN bridge can also handle the Ethernet-to-token-ring protocol conversion. Because there's a significant difference in maximum frame size between Ethernet (1,500 bytes) and token ring (approximately 5,000 bytes), the 8209 bridge uses part of the token-ring protocol to indicate to the source workstation that the maximum frame size it can use is 1,500 bytes. The smaller frame size adds overhead to the file transfer since more frames are required.

The 8209 bridge looks like a source routing bridge to token-ring workstations, while Ethernet workstations see all token-ring workstations as workstations on the same Ethernet segment. Because source routing uses redundant parallel bridge connections and spanning tree permits only a single path, the 8209 bridge permits multiple connections, but only one path can be active at any given time. The 8209 bridge operates in three different modes: token-ring-to-Ethernet version 2, token-ring-to-802.3 LANs, and a mode in which the bridge detects the type of LAN and then switches to mode 1 or mode 2.

Source Routing Transparent Bridges

A source routing transparent bridge is able to forward both spanning tree and source routing frames. This bridge uses the routing information indicator (RI) to distinguish frames using source routing from those using transparent bridging. Source routing frames set their RI indicators to 1, making it easy to distinguish this group of frames.

The movement to SRT bridges hasn't been painless. Casualties of this new type of bridge have been many of the products for bridging Ethernet and token-ring networks that performed what amounted to a protocol conversion, a transformation from one frame format to another. The IEEE 802.1D bridges don't require anything more from the workstations on the Ethernet side because Ethernet bridges handle most of the work associated with the spanning tree algorithm. Workstations on the token-ring side, however, have to construct their routing tables and build their frames to accommodate the workstations on the Ethernet side.

A Guide to Selecting Bridges

Bridge vendors are having a hard time differentiating their products from each other. The noise level and confusing jargon associated with bridges have grown to the point that selecting a bridge is a confusing, frustrating task even for knowledgeable systems integrators. In this section, I'll describe several features to look for in bridges; while you might not need specific features, this section will help you make intelligent decisions.

Packet filtering and forwarding rates

Some bridge vendors like to boast about their products' packet filtering and forwarding rates. The filtering rate in packets per second (pps) measures how quickly a bridge examines a frame, matches its address with its address table, and then decides whether to filter or forward it.

Frame sizes can vary, as can network traffic. Vendors often provide information on the number of frames per second required for various frame sizes at 50% and 100% loads. With Ethernet, 100% load conditions are unrealistic because of the number of collisions under Ethernet and the subsequent dead time on the cable. Check to see if the statistics are based on the same-sized frames.

A network manager or systems integrator must look at how a network is used on a day-to-day basis in order to make sense out of these statistics. Let's say a network uses the TCP/IP transport protocol heavily. This protocol provides unacknowledged delivery, which means that the network could become flooded with TCP/IP datagrams since source workstations don't have to wait for acknowledgment before sending large files.

Also, bridge vendors rarely specify the conditions under which their statistics are gathered. Just as the manufacturers of dot-matrix printers publish statistics on characters per second that fail to take into account the use of any special printing features, filtering and forwards rates for bridges rarely indicate the traffic conditions on both sides of the bridge.

Statistics on bridges for Ethernet-like half-duplex networks rarely point out that these statistics hold for only the condition in which the bridge is forwarding packets to a network with no current traffic to slow down the process. That's simply not realistic. Similarly, bridge statistics are usually based on the very simplest of bridges and ignore any programmed filtering, even though that might be the precise reason why the network manager selects a specific type of bridge.

Rickert (see the bibliography) has published statistics that indicate that the key to a bridge's success could be the way it handles "bursty" traffic. He believes that the size of a bridge's queue or holding area for frames is crucial. The queue should be large enough to handle reasonable temporary overloads, but not too large to introduce excessive delays. If a queue is too small, then frames will be lost when they back up in the queue. If the queue is too large, frames will remain in the queue past their "time out" and still be transmitted by a bridge that's too simple to consider such circumstances. Rickert provides some valuable formulas to help the network manager or systems integrator calculate optimal queue size.

Filtering on the basis of packet length

Some bridges can filter packets based on the actual packet length. If a network has a lot of interactive traffic (short block lengths) as well as a lot of large file transfers, a network manager might want to give priority to the interactive traffic so the response time will be faster. By programming a bridge to block longer packets during heavy traffic periods, a network manager can keep response times bearable.

"Learning" bridges

Some network managers must deal with networks where some users frequently move from area to area and other users are added or deleted on a daily basis. Some bridges require the network manager to modify the bridge's network address table each time there's a change. Other bridges are able to "learn" the locations of devices by examining the source address fields of packets they handle and then modify their own tables. These learning bridges are worth the additional expense when network managers spend inordinate amounts of time manually modifying bridge address tables.

Link ports

The ports on some bridges can be individually configured. In some cases, this means that one port can be configured for a standard 802.3 network (10Base5), while another port can be configured for thin Ethernet (10Base2). Other bridges make it possible to link a baseband Ethernet LAN with a broadband 2-Mbps Ethernet network.

Filtering broadcast and multicast packets

Some bridges have the ability to filter broadcast and multicast packets. Broadcast storms consist of a packet that's broadcast and then endlessly replicated until it creates so much traffic that it brings a network to its knees. By filtering and restricting broadcast and multicast packets, bridges can reduce broadcast storms.

Load balancing

Load balancing makes it possible for multiple ports to carry information to the same destination. By balancing the data traffic on two 56-Kbps lines, the effect is to widen the total bandwidth transmission to 112 Kbps. Different bridge vendors have their own proprietary methods of implementing load balancing. Some bridges simply divide up all traffic evenly, using a first-come-first-served approach. Others handle a specific queue first before handling a second queue. The sequencing protocols used by these bridge vendors can be important since packets might arrive out of order. Load balancing also provides a system fault tolerance feature; the built-in redundancy means that some level of communication is maintained even if one line goes down.

Bridge management and statistical software

Some bridges come with network software that provides network management features, as well as the ability to generate statistical reports on the bridge's activity. For most of its workstations, it forwards the packet across the bridge to the next network.

Most software provides information on the number of packets filtered, forwarded, refused, and rejected. Some high-end bridges offer some very sophisticated statistical reports, including network utilization and throughput, the top ten protocols used, and the top ten transmitting stations. They also provide information on which workstations transmit multicast packets. Since DEC's local area transport (LAT) protocol can only be bridged and not routed, DEC network managers need this protocol information because they can't obtain it with routers.

Summary

Bridges allow you to link subnetworks into an enterprise-wide network. By dividing large networks into smaller subnetworks and bridging them, network managers can gain greater efficiency since there's less traffic congestion. The subnets also provide greater security, because of their redundancy, than is possible with a single network. The spanning tree algorithm is used in Ethernet bridges. It requires a single path with no loops. Token-ring networks use source routing, an approach that makes the source workstation responsible for developing the complete routing path for a frame. While some Ethernet-to-token-ring bridges perform protocol conversion and transform a frame from one format to the another depending on the direction of the transmission, more sophisticated bridges are able to read a frame and transmit it in a transparent manner.

4

Local Routers and Brouters

In this chapter, you'll examine:

- How a router differs from a bridge
- Some of the major protocols associated with routing trends toward more efficient routing
- Major router features
- How a brouter operates

In chapter 3 you saw how bridges link together networks at the data link layer of the OSI model. Bridges make connected networks look like one very large network. When large networks need to be linked together and when packets need to be routed according to their higher-level protocols, a router is needed. In this chapter, you'll see how a router operates at the network layer of the OSI model and look at some of the major protocols used by routers to ensure safe delivery of packets. You'll also learn about the hybrid combination of a bridge and a router, known as a *brouter*, and investigate under which circumstances it's ideal.

What Is a Router?

A *router* operates at the network layer of the OSI model. Unlike spanning tree bridges, routers are ideally suited for large networks with several loops or redundant paths. While bridges don't concern themselves with higher-level protocols, routers are protocol-specific. A router is designed to support specific protocols, such as TCP/IP, XNS, NetBIOS, and DECnet and use the addressing schemes, error checking, and routing techniques that characterize these protocols.

Routers are particularly valuable on very large internets where the subnets might not have the same MAC layers. A NetWare SPX/IPX LAN with TCP/IP, for example, can use a router to communicate with a VAX running TCP/IP on its VMS operating system, an HP 3000 running TCP/IP on its Unix operating system, and a LAN server network running TCP/IP.

Sophisticated routers are able to route packets to other networks, even if the other networks use different addressing schemes and even different error recovery schemes.

Figure 4.1 illustrates how routers operate. Unlike many bridges, routers are able to maintain several alternative paths and select the most appropriate path given certain defined conditions, such as traffic congestion. In this case, a packet indicates a workstation on network XYZ as its destination address. Router 1 looks over its routing tables and determines that the optimum path is through router 4. Unfortunately, router 1 also realizes that the direct path to router 4 is congested. It chooses the alternative path through router 2 and router 3.

Routers are far more intelligent than bridges, and they use this intelligence to determine the optimum path for connecting together two LANs. A router is also smart enough to perform the packet segmentation and reassembly required to accommodate intermediate networks in which packet sizes are different.

Local and Remote Routers

Routers can be local or remote. This chapter examines local routers, devices that are physically connected to networks at the same site. I'll defer the discussion of remote routers to later in the book when I describe the compo-

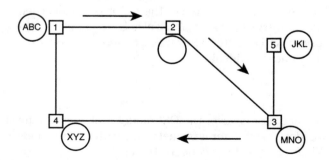

Figure 4.1 A router selecting the best path.

nents of a wide area network. At that time you'll see how routers can transmit packets over high-speed lines from one part of the country to another.

Static Routers

Static routers require the network manager to create routing tables. These tables remain static (unchanged) until the network manager changes them to reflect changes in network activity or node placement. Because static routers require manual operation to create and maintain routing tables, they aren't desirable for larger networks where routing conditions might be changing on a minute-to-minute basis.

Dynamic Routers

Dynamic routers use sophisticated algorithms to route packets along the most optimal path. Network protocol suites must have a corresponding routing protocol. Cisco Systems, for example, has a proprietary routing protocol known as interior gateway routing protocol (IGRP). IGRP considers network traffic, path reliability, and speed in selecting the optimum path.

Berkeley-derived UNIX systems use routing information protocol (RIP) to calculate how many hops though other routers different paths would encompass. They then choose the path with the fewest hops as the optimal path.

A dynamic router is constantly exchanging packets with other routers so it can learn of any new destinations or changes in existing workstations and update its address routing tables. Dynamic routers will recognize traffic congestion and failed circuits and select an alternate route if these conditions make the usual path unacceptable.

There are two major internal routing algorithms: distance vector and link state. The *distance vector* approach, sometimes also known as the Bellman-Ford protocol, is very common today and is used by RIP. The distance vector approach keeps track of the route between source and destination address in terms of hop count, or the number of routers that a packet must cross. This algorithm imposes a maximum of 15 hops for any route.

The *link state* approach requires a dynamic router to broadcast packets describing its own links to other routers. All routers on the network use these broadcast packets to assemble their own routing tables. You'll return to this topic later in this chapter when we examine a routing protocol that's growing in popularity, known as OSPF.

Routers Create "Firewalls"

One major advantage a router has over a bridge is that it doesn't automatically replicate all broadcast messages. This means that if a device begins to flood a network with copies of a single packet, the routers are able to keep the problem local by presenting a "firewall" that prevents the storm from engulfing the entire network.

Routers Make Management of a Large Internet Much Easier

Routers can take advantage of addressing schemes such as the one used by internet protocol (IP) to create subnetworks. An IP address includes a network number, subnetwork number, and host number. XYZ Corporation, for instance, might have a corporate headquarters as well as three local plants, each of which has its own subnetwork. The host number refers to any IP device that has an IP address; these devices can take the form of bridges, personal computers, mainframe hosts, etc.

Figure 4.2 illustrates a large company that has four major networks. The VAX handles accounting while other departments use a Banyan server running Vines, a server running Microsoft NT, and a Novell file server running NetWare. Rather than concern itself with the incompatibilities of the various network operating systems, the company opted to run TCP/IP on all four networks and then link them together via routers.

Routers can be programmed to be very selective. Note also that routers are redundant, which means that if one path is blocked, packets can be routed via an alternate path. Also, routers make network management easier by offering network management software that monitors and controls network operations. The TCP/IP routers used in the previous example provide simple network management protocol (SNMP), a topic you'll return to in chapter 11 when you look at network management systems.

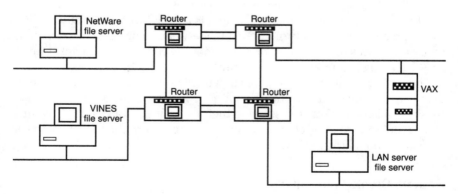

Figure 4.2 Routers linking subnets together for more efficient management.

Point-to-Point Protocol (PPP)
for Multivendor Router Communications

There are many examples in data communications of standards that aren't very standard. The RS-232C standard is a good example; printer manufacturers have never agreed on which pins should be used for specific handshaking functions. As a result, you can't simply substitute one printer's serial cable for a different vendor's printer cable and expect it to work.

The same sort of problem exists with routers. While TCP/IP routers use the internet protocol for routing their datagrams, vendors use serial line internet protocol (SLIP) as a basis for developing different ways to encapsulate IP datagrams. This has made it difficult or even impossible to mix and match TCP/IP routers on the same network. Lack of router compatibility has been a particular problem for enterprise networks in which, for instance, one plant uses a low-end router to satisfy its simple needs and another facility uses a sophisticated router from another vendor to meet its complex needs. The Internet Engineering Task Force has developed the point-to-point protocol (PPP) for this purpose.

PPP replaces SLIP with a standard method for encapsulating IP datagrams. This new protocol means that systems integrators can design direct serial connections between TCP/IP routers at very high speeds ranging from 9.6 Kbps for dial-up lines to T-1 and fractional T-1 service.

Open Shortest Path First (OSPF)

The U.S. Department of Defense's Internet Activities Board sets internet policy for TCP/IP users. The board's Internet Engineering Task Force (IETF) has created the Open Shortest Path First (OSPF) working group to develop a dynamic routing protocol for TCP/IP that will provide features not offered by routing internet protocol (RIP).

RIP has limitations when used with networks of more than 100 routers because of RIP's reliance on the Bellman-Ford algorithm. This approach requires the frequent broadcast of the entire routing table. On large internets with over 100 routers, routing updates take longer and longer and consume increasing amounts of bandwidth.

RIP has other limitations. Packets can't travel through more than 15 routers from sender to receiver. This protocol selects a single path to each destination and isn't capable of considering factors such as traffic congestion, delay, and bandwidth.

The OSPF protocol uses a link-state and shortest-path-first algorithm. Each router broadcasts a packet that describes its own local links. Routers collect information from these broadcast packets to build their own network routing tables. Since these packets describing local links are very short, they cause far less traffic congestion than RIP's approach of broad-casting very large routing tables describing the entire network.

Another advantage of OSPF is that network managers can configure their routers to provide least-cost routing according to whatever criteria these managers define as a "cost." Unlike RIP, OSPF doesn't limit the number of routers that can be used, nor does it limit routing to a single path; loads can be distributed over several different paths to optimize available bandwidth.

OSPF provides far more flexibility than RIP when it comes to type of service. This new protocol offers eight classes of service with separate paths available for each path. Network managers can program their routers so certain types of packets (large file transfers that aren't time-sensitive, for example) are sent via satellite with delays that could stretch to several hours. Time-sensitive packets, on the other hand, could be given a class of service to route them via more expensive phone lines.

The good news about OSPF is that the task force has developed procedures so large networks can run both RIP and OSPF as a dual protocol stack, for a temporary solution while they work on converting to the latter protocol. OSPF can route information to RIP transparently so users aren't even aware of the conversion process.

Features to Consider When Purchasing a Router

It's a jungle out there. There are dozens of routers, and each one claims to be superior. Whether a particular feature is really beneficial or not is a decision you'll have to make as network manager or as the systems integrator designing an entire project. In this section, I'll explain some of the more significant features routers offer so you can make an informed decision.

Number and type of local and wide area network interfaces

The number of interfaces available on a router vary widely according to vendor and model. A high-end router might offer seven or more LAN ports and 14 or more WAN ports. Since routers are protocol-specific, a high-end router might support 15 or more protocols, including Ethernet versions 1 and 2, IEEE 802.3, IEEE 802.5, and several proprietary network protocols. The reason why multiple WAN ports are so important is because WAN links are much slower than local links and the bandwidth is much smaller.

The type of WAN interface on a router is just as important as the number of ports available. A network manager or systems integrator must select the appropriate router model for a specific network design. While many models offer RS-232C, RS-449, and CCITT V.35 interfaces, a few also offer a fiber-optic FDDI interface, and even an X.25 interface to public data networks.

Network management protocols supported

On large networks, it's essential to be able to gather detailed routing statistical reports as well as fine-tune routers for optimal performance. Routers vary widely as far as the network management software they support. Some support the IEEE 802.1 network management standard, others support simple network management protocol (SNMP), which will be described in chapter 11's discussion of network management. Some routers support only the vendor's own proprietary network management software. Systems integrators must consider long and hard whether or not they want to be locked into a specific proprietary network management scheme that will probably remain static and not grow the same way as industry supported standards.

Router performance

The number of packets per second (pps) a router can handle is very revealing. There's a lot of overhead involved in routing decisions. The pps figure takes into account the time required for routers to access their tables and decide on the optimum path. Unfortunately, every router vendor has a different way of measuring pps. Do the packets travel in both directions? Do the packets not have to be segmented to travel to a network with a different packet size? How many packets are lost? What happens to speed when the router is programmed to permit only certain types of packets to pass?

Protocols supported

Because there are multiprotocol routers available today, it's essential to consider future as well as present routing needs. TCP/IP has different broadcast formats; will the router's protocol implementation be able to support the different TCP/IP versions on an enterprise network? Similarly, while TCP/IP and its accompanying SNMP management protocol might be acceptable for the present, does the router also support the OSI suite of protocols and its CMOT management protocol? OSI protocols that could prove essential in the future for network routing include connectionless network protocol (CLNP), end system to intermediate system (ES to IS) routing, and network service access point (NSAP).

What types of networks will join the enterprise network in the future? If UNIX systems are on the drawing board, then routing information protocol (RIP) is essential since it's the interior routing protocol used on Berkeley-derived UNIX systems. Will there be communication in the future with the Defense Data Network (DDN)? DDN supports exterior gateway protocol (EGP), which is also known as request for comment (RFC) 888 and 904. Security requirements on the Department of Defense network require an IP router that can provide datagrams that support connection to the Blacker interface for secure public data networks (X.25). Table 4.1 lists some of the

**TABLE 4.1 Major Protocols
Supported by Routers**

Protocol	Source
IP	RFC 791, 1009
RIP-IP	RFC 1058
TCP	RFC 793
SNMP	RFC 1065, 1066, 1098
CMOT	RFC 1095
IPX	Novell
XNS	XSIS028112
RTMP	Apple
NBP	Apple
EP	Apple
ZIP	Apple

major protocols routers support, along with the standard that defines them where appropriate.

Security

Some routers offer security options that enable you to filter out packets bound for proprietary or secured systems on the basis of their IP addresses. These routers can also be programmed to filter on the basis of message type, which means you can permit electronic mail and ban file transfers.

Bridges Versus Routers

While chapter 3 was devoted entirely to bridges, this chapter has dealt only with routers. One major issue confronting network managers and systems integrators is distinguishing the need for a bridge from the need for a router. In this section, I'll compare and contrast the two types of network interoperability devices.

Bridges are ideal when two networks with different higher-level protocols but the same MAC layers need to be linked together. Bridges are relatively inexpensive and much faster than most routers. They're also much easier to install and maintain. Once installed, bridges can automatically learn the network location of stations by listening to the source addresses of network traffic.

Bridges are not ideal, though, for large, complex networks for a variety of reasons. Since bridges pass all traffic, including broadcast storms, a few NIC problems could bring down a very large internet. Also, since many bridges require a single path between networks, they lack the system fault tolerance that routers' multiple paths provide.

More and more networks are now running multiple protocols, so a major advantage of a router is its ability to pass packets with specific protocols from one network to another. Routers using dynamic routing schemes are able to adjust to changing network conditions and provide network management functions not offered by bridges.

Another major advantage of a router over a bridge is its ability to perform packet segmentation and reassembly in order to accommodate intermediate networks in which packet sizes are different. An example of this situation would be connecting two Ethernet networks running NetWare via an Arcnet network running NetWare. Ethernet and Arcnet packet sizes vary considerably in size.

While they're considerably more complex and more expensive than bridges or routers, some situations might require a hybrid of the two devices, a brouter. The next section examines this new network tool.

Brouters

While definitions vary widely by vendor, in this book I'll define a *brouter* as a hybrid bridge and router that's able to perform both functions. It first attempts to make a routing decision, but reverts to bridge status if unable to do so.

In Figure 4.3, an IBM 9370 host uses the IBM source routing approach to communicate with an IBM Token-Ring Network. The Token-Ring Network is connected to another 802.5 LAN, this time with a Novell NetWare file server using the IPX protocol; the brouter uses 802.5 MAC layer procedures to pass its data transparently between these two token-ring networks.

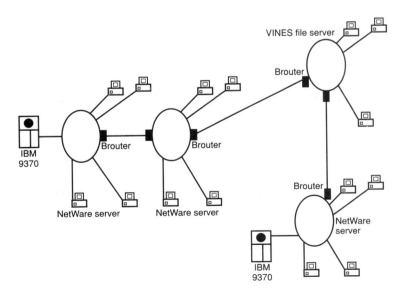

Figure 4.3 A brouter in operation.

The NetWare token-ring network doesn't use source routing. The brouter is able to recognize source-routed frames and forward them via the route defined in the frame. If the packet doesn't have its source routing information, the brouter can provide the best route to the destination address. In a mixed environment, this particular brouter can coexist with other vendors' source and transparent bridges as long as it's the first and last bridge in the chain, as pictured in Figure 4.3. Finally, the packet is then routed to an IBM Token-Ring Network that does use source routing.

How a brouter uses its routing tables

Figure 4.4 illustrates a brouter in action; I'll use RAD Network Devices' Extended Ethernet LAN to illustrate how brouters use their routing tables. This brouter is attached to an Ethernet or IEEE 802.3 LAN like any other node. Serial links connect the brouters.

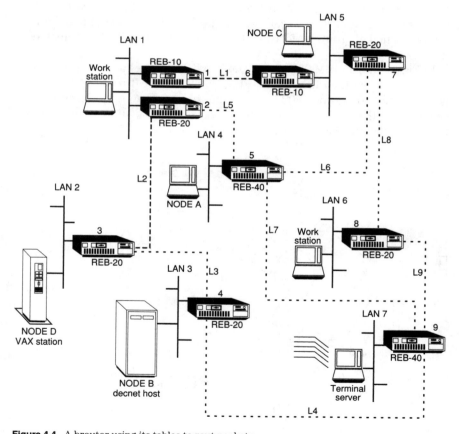

Figure 4.4 A brouter using its tables to rout packets.

RAD's brouters use a database called LAN table to store addresses for nodes attached to their own LAN. If an address isn't detected after a certain period of time, it's deleted from this table. A second database, called net table, contains all the node addresses for the extended network in terms of particular bridges. A third database, called routing table, contains directions for the optimal and second-best paths for routing packets to each bridge in the network. These brouters broadcast messages periodically that update modified.

Let's assume that node A wants to send a packet to node B. The following steps (courtesy of RAD Network Devices) would take place:

1. Bridge 5 uses its LAN table and net table to conclude that the packet has to be forwarded to a LAN connected to bridge 4.

2. Bridge 5 uses its routing table to find out that the best path to bridge 4 is via L7. The packet is transmitted to L7.

3. Bridge 9 uses its routing table to determine the best way to route the packet. It sends it to L4.

4. Bridge 4 receives the packet, deencapsulates it, and then sends it to LAN3.

5. If L4 fails, bridge 9 is aware of this fact and uses its routing table to send the packet back to L7, which is the second-best path to bridge 4.

6. Bridge 5 receives the same packet it transmitted and forwards the packet using the second-best direction, i.e., to L5.

7. Bridge 2 to L2, bridge 3 to L3, and bridge 4 to LAN 3 completes the sequence.

Summary

Routers function at the network layer of the OSI model and are protocol-specific. One function they perform is the creation of "firewalls" that prevent broadcast storms on one network from sweeping across the entire internet. Routers can be programmed to provide different classes of service and to use different routes for different types of packets.

Industry committees have developed new standards, such as point-to-point protocol (PPP) and open shortest path first (OSPF), that will make routing much more efficient. While routers differ widely when it comes to features, routing speed, protocols available, programmability, and security are important criteria when evaluating routing products.

Brouters are hybrid bridges/routers. They can route certain specific protocols and then provide bridging for all other protocols. This versatility makes them very desirable on large internets.

5

Wireless Networks

In this chapter, you'll examine:

- The different types of wireless LANs
- How frequency hopping works
- How hybrid wireless and wired LANs are designed
- Alternative approaches toward wireless wide area networks

Network managers have new tools for linking their networks together, in the form of wireless technology. Wireless LAN segments can be particularly useful in environments where cabling is difficult to implement, such as isolated warehouses, reception areas, and departments where workgroups change frequently. While this technology is still in its infancy and contains a number of limitations, including bandwidth, it's developing rapidly and will become even more popular when prices drop.

Today's wireless LANs use one of three transmission technologies: infrared, spread spectrum, or narrowband radio. Figure 5.1 provides an overview of available bandwidth for a wide variety of transmissions. The discussion that follows explains the advantages and disadvantages of each major type of wireless LAN transmission technology.

Infrared LANs

Wireless infrared LANs offer a number of advantages to users. They can match the speed of wired LANs, including the 16-Mbps speed offered by to-

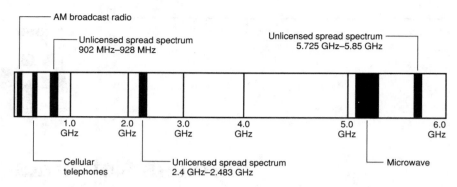

Figure 5.1 The electromagnetic spectrum.

ken rings. Their price/performance ratio is much more favorable than that of spread spectrum LANs. A third benefit is that infrared signals offer greater security than spread spectrum radio signals.

The major limitation of this technology is that it requires a clear line of sight between transmitters and receivers. When infrared LANs are installed outdoors to perform tasks such as connecting buildings across a parking lot, their signals are susceptible to interruption during adverse weather conditions. Finally, infrared LANs can transmit only over limited distances, up to approximately 100 feet.

In early 1994 the Infrared Data Association introduced the infrared coder/decoder standard. This set of specifications ensures that by 1995 there will be interoperability among infrared products developed for mobile computers. Apple has already announced support for this new standard. Table 5.1 summarizes some of the major strengths and weaknesses of this technology. There are three distinct types of infrared LANs: line-of-sight, scatter, and reflective systems.

Line-of-sight infrared LANs

The line-of-sight variety is limited to locations such as offices where there are no physical obstructions between users' workstations. Although the point-to-point nature of this technology restricts distance to around 100 feet, transmission speed can match that found in cabled networks. Infralink's InfraLAN product family, for example, now provides both 4-Mbps and 16-Mbps token-ring infrared LANs.

Scatter infrared LANs

Scatter infrared technology bounces signals off walls and ceilings to "illuminate" an area of approximately 100 feet. This approach produces a relatively low-speed signal.

TABLE 5.1 Advantages and Disadvantages of Infrared Technology

Advantages	Disadvantages
Wired LAN transmission speeds	Susceptible to interruption from adverse weather conditions
Can use off-the-shelf network interface cards	Requires line of sight

Reflective infrared LANs

In reflective systems, optical transceivers mounted near PC workstations are directed toward a common spot on an office ceiling. This approach works well in an environment with high ceilings. Photonics' Photolink is an example of this type of technology.

Spread Spectrum Radio LANs

Developed by the military to provide reliable, secure transmission, spread spectrum technology uses one of two different techniques to "spread" data over several frequencies. The direct sequence code division multiple access (direct sequence CDMA) approach represents each data bit by a bit pattern called a "chip." 1s and 0s are represented by chips that are inverses of each other. Each bit is spread over a wide-frequency spectrum when transmitted. The receiver collapses each chip back into a single bit. Because all signals that don't match are eliminated, the resulting signal is free of any interference. The alternative spread spectrum approach, known as frequency hopping CDMA, changes the frequency of each bit transmitted according to a predetermined pattern.

In the U.S., an 83-MHz-wide channel is available from 2.4000 to 2.4835 GHz for unlicensed operation. This frequency band is divided into 82 1-MHz channels, or *hops*. Adapters tune in on a specific hop for a short period of time and then move to a different hop. The hops are visited in a predefined order, as specified by the hopping sequence. All stations participating in the same network use the same hopping sequence and synchronize their hop timing. When interference is present, it usually affects only a few hops. Hopping sequences are designed so that successive hops are usually several MHz apart.

The FCC authorized that three separate frequency bands be set aside for commercial radio-based LANs: 902 MHz to 928 MHz, 2.4 GHz to 2.5 GHz, and 5.8 GHz to 5.9 GHz. Many of the first wireless network products were designed to work in the 902- to 928-MHz unlicensed industrial, scientific, and medical (ISM) band. In 1993, the trend changed as more companies began offering products in the other two higher-range ISM bands. The higher the frequency selected, the more expensive the components. No matter which approach is used, the result is the same.

TABLE 5.2 Advantages and Disadvantages of Spread Spectrum Radio Technology

Advantages	Disadvantages
No FCC license required	Limited distance (1-watt power limit of around 800 feet)
High immunity to interference	Frequency conflicts posssible with multiple LANs in a high-rise environment
No line of sight required	

Spread spectrum technology can transmit its low-frequency signals through common building materials and thus has a greater range than other wireless LAN transmission alternatives. While virtual immunity to signal interference and significant security complement spreads spectrum's extensive range, the major limitation of this technology has been that its considerable bandwidth consumption results in a relatively low transmission speed compared to other wireless technologies. Table 5.2 summarizes the major advantages and disadvantages of this technology.

Interference problems with spread spectrum technology

Reports from customers using spread spectrum technology indicate that, despite the theory, there are some problems with interference. Effective throughput drops when two pairs of spread spectrum products are installed side by side. It's unlikely that several spread spectrum devices can be installed at the same site without seriously degrading throughput. Other forms of electromagnetic emissions also affect spread spectrum throughput rates. Heavy machinery will degrade throughput as, apparently, will nearby microwave ovens.

European and Japanese spread spectrum technology

While U.S. vendors chafe over restrictions imposed by the FCC, European and Japanese wireless vendors also suffer from limitations imposed by their own regulatory bodies. The European Telecommunications Standards Institute has adopted the Digital European Cordless Telecommunications specification known as DECT, which is supported by the EEC Directive 91/287. DECT defines a spread spectrum band for the European Community between 1.88 and 1.90 GHz. This 20-MHz bandwidth will probably provide throughput of around 1.1 Mbps. The indoor range is limited to a maximum of around 100 feet. An ETSI subcommittee is developing a 10-Mbps standard based on 150 MHz of bandwidth in the 5-GHz range. This specification is expected to be completed in 1995.

Japan's wireless technology is controlled by the Japanese Ministry of Posts and Telecommunications (MPT). Its Telecommunications Technology Council is developing specifications for spread spectrum LANs. An industry group funded by Japanese electronics companies is investigating wireless LANs in the 2.4- to 2.5-GHz and 180- to 190-GHz ranges. Such bandwidth would support 10-Mbps wireless LANs. Japan and Europe trail the U.S. in the manufacture of wireless LANs.

Narrowband Radio LANs

Narrowband radio LANs use dedicated, licensed bandwidth at 18 to 19 GHz, which is assigned by the FCC. Because narrowband radio is available in bandwidth greater than in the spread spectrum range, transmission rates are generally higher than those found in spread spectrum technology. The signal cannot penetrate metal walls or concrete bearing walls within a building, but its high frequency enables it to have a range of 5,000 square feet. Motorola's Altair is the major product that's currently available using this technology. Motorola has licensed this bandwidth in most metropolitan areas and reallocates it to customers to spare them the difficulty of dealing with FCC paperwork.

The IEEE 802.11 Committee's Wireless LAN Specifications

The IEEE 802.11 committee has been developing specifications for wireless LANs that will support peer-to-peer, ad-hoc, and infrastructure LANs interconnected via an access point with an existing wired network. The protocols being developed will permit the mobile user to roam freely throughout a facility or campus while maintaining seamless connection to network resources. They'll also permit power conservation, which will enable small, mobile computing devices to communicate for long periods of time on a single battery charge.

The committee is defining protocols for wireless LANs in the 900-MHz, 2.4-GHz, and infrared frequency bands. It's also defining different physical (PHY) specifications for medium-dependent protocols. There are different PHY specifications for each frequency band supported in the 802.11 standard. There are medium access control (MAC) specifications for ad-hoc wireless networks and wireless network infrastructures. A single medium-independent MAC protocol provides a unified network interface between different wireless PHYs and wired networks.

User requirements for a wireless LAN established by the 802.11 committee in March 1992 included support for overall bandwidths of 1 Mbps or more in the following application areas:

- File transfer and remote control
- Program loading
- Program paging

- Transaction processing
- Multimedia
- Process control
- Manufacturing monitoring and control
- Material handling
- Patient care systems

Roaming Wireless LANs

Xircom has coined the term *cordless LAN* to describe a wireless local area network in contrast to a wireless wide area network. One of the major uses of a wireless or cordless LAN within a building is to enable employees to communicate while they're in meetings. As these people move from one area to another area within a building, they can remain on the network. Xircom has developed protocols that enable "roaming" users to remain on a wireless network. Other companies, such as Proxim, now also offer this feature.

Bridging Buildings to Connect Wireless LANs

There are several spread spectrum LAN bridges that can connect buildings up to three miles apart at speeds up to 2 Mbps. Solectek's AirLAN/Bridge is an example of this type of device. Transmitting in the 902- to 928-MHz band, this device supports both Ethernet and token-ring networks and is compatible with most network operating systems including NetWare, LAN Manager, Pathworks, and Vines. At 2 Mbps, this device can transmit e-mail between buildings, but would prove impractical for large data file transfers.

Factors Delaying Implementation of Wireless LANs

There are a number of reasons why wireless LANs haven't yet become very popular. The high cost of equipment coupled with the rapidly declining prices of cabled Ethernet adapters has widened rather than reduced the price differential between the two technologies, and made it even more difficult to justify opting for wireless LANs.

A second reason why wireless LANs haven't taken off in popularity is that standards have been slow to develop. The IEEE 802.11 committee has taken its time finalizing specifications, and corporations have been reluctant to invest in products that could wind up being proprietary.

Wide Area Wireless Networks

The growth of wide area wireless networks has far outstripped that of local area networks. In this area there has also been a battle over standards. The result is that network managers have to make some tough decisions in order to avoid being stuck with what could become proprietary technology.

Cellular Digital Packet Data (CDPD)

Cellular digital packet data (CDPD) is a joint venture that includes McCaw Cellular Communications and the regional Bell companies that provides "channel hopping," the ability for data transmissions to occur on existing analog voice networks during idle times between voice calls, at speeds up to 19.2 Kbps. Encrypted packets sent over CDPD include a user identification that makes it hard to scan and decode the messages. By early 1995 CDPD should be a real force in cellular data transmission.

While the packet-switched competitors to CDPD suffer several-second delays that make them unsuitable for two-way interactive communications in real time, CDPD should only have subsecond delays. There should be approximately 10,000 CDPD cell sites when the network is fully developed.

Packet-Switched Wireless Data Networks

Packet-switched wireless data networks transmit standard amounts of data, packaged with an address. Each node checks the packet's address and then rebroadcasts it to the next node. The receiving node then reassembles the packet. Messages from all over can share the same frequency.

The Ardis system

The 19.2-Kbps, 800-MHz Ardis system is a joint venture of IBM and Motorola. Designed originally to support field service personnel, it uses a packet-switching technology to carry data transmission. Ardis has approximately 1,300 base stations serving around 400 metropolitan areas.

RAM Mobile Data

RAM Mobile Data was formed by a partnership between Bell South and RAM Broadcasting Corporation, using equipment by L.M. Ericsson. RAM provides Mobitex networks throughout the U.S. and Europe and users are required to have a Mobitex modem, which uses proprietary protocols to transmit and receive information.

The company holds FCC licenses for between 10 and 30 frequency channels in each of the top metropolitan areas across the United States. RAM has around 800 base stations serving around 100 metropolitan areas, and has been very aggressive in signing up software developers to create applications for this network. Intel has agreed to use Ericsson Mobidem hardware and the RAM Mobitex standard in their handheld products.

The company has also created a wireless messaging business unit to create systems that link its network to wired computer mail systems. RAM Mobile Data's service is available in over 6,300 cities and towns, covering 90 percent of the U.S. urban business population.

Novell's vision of a LAN client

Because around 70 percent of corporate LAN users are found on LANs running Novell's NetWare network operating system, it's crucial that any major wireless technology be consistent with the plans of this Provo-based company. Novell's vision of a wireless LAN client includes software for remote LAN access as well as the integration of WordPerfect Office. The company plans to add wireless support to NetWare in order to support wireless LANs and give them access to NetWare over wireless messaging and paging networks. The wireless client will be based on Novell's VLM client, developed for NetWare 4.*x*, and will include a version of its IPX protocol optimized for wireless systems calls, queries, and data. Novell is also working with Motorola to support access to NetWare from various cellular messaging and paging networks. A NetWare loadable module gateway to Motorola's network integration (MNI) service is being developed. MNI integrates the Ardis and RAM Mobile Data networks and EMBARC and SkyPage messaging services.

Hybrid Wireless and Wired Networks

A few years ago, AT&T introduced its spread spectrum wireless LAN, WaveLAN, as a wireless replacement for cabled LANs; it failed. The pricing differential between wireless and cabled equipment was, and still is, too great to justify a large wireless LAN, except where conditions make it impossible to use conventional cabling. Today, AT&T markets the product as a hybrid solution for environments difficult to wire that need to be connected to wired LANs. Figure 5.2 illustrates a conventionally cabled Ethernet LAN segment with a server containing a wireless adapter card that bridges this segment to a server on a wireless LAN segment.

Summary

As companies move toward enterprise networks that incorporate all parts of a company, wireless technology is going to play a more significant role. For the next few years, this role is probably going to be a part of a hybrid network, with the wireless segments providing network functionality where conventional cabling isn't easy to implement. Wireless technology is bound to play a much more significant role on wide area networks before wireless LANs take off.

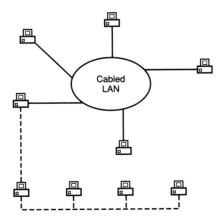

Figure 5.2 A hybrid network containing wireless and wired segments.

WaveLAN wireless LAN

Infrared technology offers transmission speeds consistent with those offered by cabled networks; unfortunately, distances are limited and network users must often be situated with a line of sight for this type of network to function properly. Spread spectrum technology offers less transmission speed (around 2 Mbps), but longer distances and, at this time, far more choices in vendors. The roaming technique pioneered by Xircom permits users to move throughout a building while their signals are passed from access station to access station.

6

Gateways to the Mainframe World

In this chapter, you'll examine:

- Several different ways a LAN can have a gateway to the IBM mainframe world
- How IBM's systems network architecture (SNA) is changing to make it easier for peer-to-peer communications with PCs on a LAN

One crucial issue facing network managers is the best way to link their LANs to corporate mainframes. In this chapter, I'll describe several different types of LAN gateways, including a relatively new direct connection, and also remote gateways.

In order to understand what type of gateway you need, it's helpful to review the building blocks of an IBM mainframe system, including systems network architecture (SNA). This chapter will look at SNA as it is today as well at some new programming interfaces that promise greater direct communication between mainframe and LAN programs in the future. In other words, the topic of this chapter is gateways from the LAN world, populated with microcomputers, to the world of mainframes. It's a complex subject, particularly for LAN managers with little mainframe experience, so the chapter assumes no prior knowledge.

The IBM Mainframe World

Before we examine the several different ways LANs can be linked to main-frame computers, it's worthwhile to spend some time looking at the main-frame environment, particularly because many of you might be far more comfortable with micros and LANs than with host computer terminology.

I'll focus initially on the IBM mainframe world since IBM controls such a significant market share. Figure 6.1 illustrates the IBM mainframe world; it's a world of centralized processing dominated by the host computer. For the most part, this is a world in which the mainframe has master/slave rela-tionships with its peripherals, which means that the host polls its various network devices to see if they want to communicate. You'll return later in this chapter to the growing movement toward peer-to-peer relationships in which devices communicate with each other when needed as well as with the host.

Terminals and other devices communicate via *cluster controllers*, de-vices that serve as an interface between the host computer and its many network devices. The high-end 3175 cluster controller can support a theo-retical maximum of 253 devices. Terminals dialing from remote sites via mo-dem communicate with a communications controller, a high-speed "traffic cop" that transmits this information to the mainframe.

Figure 6.1 The IBM 370/168 mainframe environment (*IBM*).

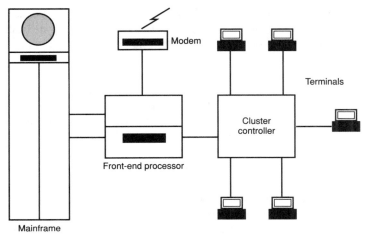

Figure 6.2 The basic components of a mainframe system.

Courtesy of IBM

The name of the game in mainframe operations is offloading as much work as possible to other processors and leave the host free to do what it does best—crunch numbers very quickly. A *front-end processor (FEP)* is a software-programmable controller that relieves the host of many networking and data communications tasks. It can handle such "overhead" functions as polling devices, error checking and recovery, character code translation, and dynamic buffer control.

IBM front-end processors are usually referred to by their model numbers, the most common of which are the older 3705 and newer 3725. On an IBM mainframe network, the network control program (NCP) is generated on and loaded from the host onto a front-end processor. Software on the host, called virtual telecommunications access method (VTAM), communicates at a speed of approximately 2.5 Mbps with the NCP running on the FEP. The FEP also concentrates several low-speed data transmissions into a steady, high-speed flow of information to the host.

Generally, terminals and printers aren't connected directly to the FEP, but are connected in clusters to control units, known as *cluster controllers*. In the IBM mainframe world, these devices are also known by their model numbers, which most often include the numbers 3174, 3274, and 3276. Figure 6.2 illustrates how these various IBM devices are linked together in a typical corporate environment.

The level of a gateway's functionality can vary

A *gateway* is a device that connects networks with different network architectures. Gateways use all seven layers of the OSI model and perform the protocol conversion function at the application layer. Typical corporate gateways connect the PC world of token-ring, Ethernet, and AppleTalk LANs to IBM's mainframe SNA environment with X.25 packet-switched networks or DECnet networks.

LAN gateways differ considerably in their functionality. At the very lowest level, a gateway provides terminal emulation so all LAN workstations can emulate or imitate a dumb terminal. Even in this case, the level of emulation can vary considerably depending on the gateway. Some gateways handle IBM's block-data transmission approach very well and map PC keyboards so they have the look and feel of an IBM 3270 terminal, while retaining the advantage of an intelligent PC workstation by permitting easy switching from terminal emulation to PC operations with a hotkey. Some gateways permit a LAN workstation to window several different host sessions and move easily from session to session.

A second level of LAN gateway functionality includes file sharing between LAN and host. Novell has developed a platform-independent version of NetWare that will run on several different platforms, including several traditional minicomputer platforms. NetWare for VMS is already available, so it's now possible for a NetWare LAN user to log onto a VAX from a PC workstation and view and access NetWare files residing on the VAX. To the LAN user, the VAX appears to be just another PC file server. Since the VAX is running NetWare as one of the multitasking processes under VMS, a VAX user could log onto the computer and exchange information with the VAX concerning these NetWare files using standard DEC commands. In other words, the NetWare and DECnet protocols are able to communicate with each other in a manner that's transparent to the end user. In a similar way, with Novell's UNIXWare running on an application server, you can view UNIX files as well as NetWare files and handle both these different file formats.

At a still higher level of functionality, a gateway could provide peer-to-peer communications between microcomputer programs running on the LAN and mainframe programs running on the host. These types of client/server relationships will become more and more important in the near future as programs are written to distribute databases among LANs, minicomputers, and mainframes, with the machines' users communicating with the programs using the same type of user interface.

Later in this chapter, I'll describe IBM's systems application architecture (SAA), its set of specifications for providing a uniform user interface and common communication links among its whole family of computers. At this point, it's worth noting that NetWare for SAA already supports IBM's SAA services. This means that, when programs adhering to SAA are written, Novell's NetWare will be able to facilitate the type of communications required for programs to exchange meaningful information.

How do gateways link hosts and LANs?

Before LAN gateways, companies connected PCs directly to IBM front-end processors via coaxial cabling using expensive PC 3270 emulation cards and software. Many companies invested tens or even hundreds of thousands of dollars in products providing a single connection. With a LAN gateway, however, the micro-mainframe connection is much more cost-effective. The gateway board emulates a cluster controller so each network workstation is seen by the mainframe as a terminal linked to the cluster controller. The gateway's multiple mainframe sessions are split among the network's workstations, so the channel rarely sits idle. Only the gateway need have a circuit card and the software necessary for protocol conversion and terminal emulation.

Figure 6.3 illustrates four different types of LAN connections to an IBM mainframe: a local connection between a host and a LAN gateway workstation via coaxial cable, a local connection between a LAN and front-end

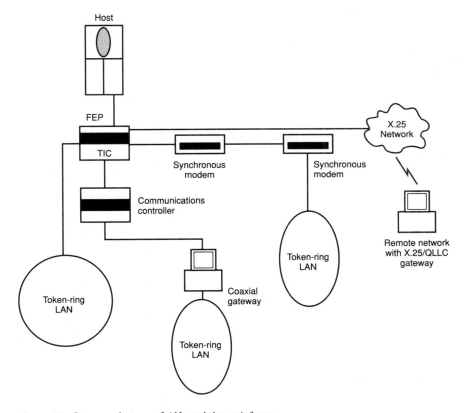

Figure 6.3   Gateways between LANs and the mainframe.

processor via a token-ring interface coupler (TIC) gateway, a LAN remotely linked to a host's front-end processor via modems, and a LAN remotely linked to a mainframe's front-end processor via an X.25 network.

Local DFT coaxial gateway

With earlier IBM systems, when the host computer requested transmission of a terminal's contents, the controller and not the terminal handled the request. This approach was known as the control unit terminal (CUT) mode of operation. IBM's 3174 family of controllers introduced the concept of off-loading this terminal processing task to programmable terminals, which became known as the distributed function terminal (DFT) mode of processing. A DFT coaxial gateway allows a LAN workstation to emulate a distributed function terminal.

The gateway PC acts like an SNA cluster controller communicating with its workstations by emulating DFT terminals. Workstations on the LAN run their own software that allows them to emulate an IBM terminal. The workstations' terminal-emulation software allows them to access the gateway via the LAN's transport protocol. The number of concurrent sessions available with these gateways varies widely, depending on the manufacturer. While some low-end gateways permit as few as five concurrent sessions, other gateways can handle more than 40. It's possible to install multiple coaxial gateways when traffic is too heavy for one to handle efficiently.

The token-ring interface coupler (TIC) gateway

The highest-performance LAN gateway is the link between a token-ring network and a host's FEP, via a token-ring interface coupler (TIC) gateway. The TIC permits a 4-Mbps or 16-Mbps connection, depending on the hardware installed. The key component for this connection is the Token-Ring Adapter, not the NIC, but an IBM FEP hardware upgrade. An IBM 3745 Token Ring Adapter comes with two 802.5 connections or TICs. High-end FEPs offer additional token-ring network connections. The 3745 model 130 LAN gateway offers four, while the 3745 models 210 and 410 each offer eight connections. A non-token-ring LAN, such as Ethernet, can be linked via TIC gateway at speeds of 2.35 Mbps.

The gateway PC is viewed by the mainframe as a cluster controller; the mainframe polls this gateway PC while it in turn polls all other workstations on the token-ring network.

Remote LAN gateway

As enterprise networks and wide area networks evolve, remote LAN gateways are becoming very common. A PC on the remote site's LAN functions as a gateway and runs gateway software. This gateway PC functions as a cluster controller and communicates with a front-end processor using IBM's synchronous data link control (SDLC) protocol via synchronous modems located at both sites. Figure 6.4 illustrates a typical remote LAN gateway.

The limitation of remote gateways has traditionally been speed. A synchronous modem can dial up a front-end processor at speeds up to 64 Kbps. Companies with heavy micro-mainframe traffic might require multiple remote gateways to solve this congestion problem.

X.25 gateways

Remote LANs can also communicate with IBM mainframes via X.25 gateways. A gateway PC with an adapter card functions as a cluster controller and runs special gateway software that contains the QLLC protocol, an IBM-defined protocol that runs over the X.25 suite. The other LAN workstations emulate IBM 3270 terminals. The IBM host simply assumes it's communicating with a remote cluster controller.

Gateways linking together LANs

In enterprise networking, it's possible for several LANs to be linked to a company's mainframe but not to each other. When they need to be linked together, a mainframe can act as a router to connect the LANs. Software has been developed for NetWare and LAN Manager that enables these LAN

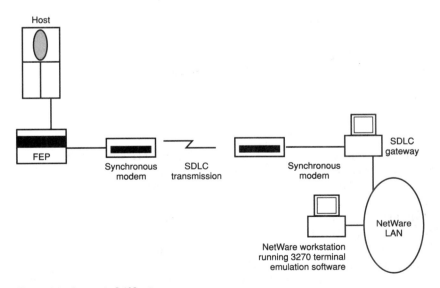

Figure 6.4   A remote LAN gateway.

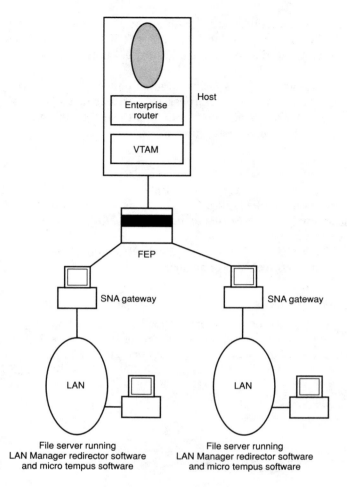

Figure 6.5 Routing LAN information through a mainframe.

operating systems to run as an application under IBM's virtual telecommunications access method (VTAM) software on mainframes running VM or
MVS. As Figure 6.5 illustrates, a NetWare LAN uses its SNA gateway to send
information to a mainframe. A NetWare application program running on the
host under VTAM sends this information via a different gateway to its destination, a workstation on a second NetWare LAN.

LAN traffic can piggyback on top of the LAN-to-host communications and
eliminate a company's need to purchase expensive routers and bridges for
remote communications as part of a wide area network.

Systems Network Architecture (SNA)

Systems network architecture (SNA) is IBM's network architecture, its suite of layered protocols designed to facilitate data communications for an IBM computer system environment. IBM announced SNA in 1974 to provide host to terminal connections under its synchronous data link control (SDLC) protocol. Information was to be stored on the mainframe using virtual telecommunications access method (VTAM), while communications between the host and attached devices would take place via the network control program (NCP) running on a front-end processor. In this "master/slave" world, the host periodically polled the dumb terminals to see if they had information to transmit.

Network addressable units (NAUs)

Every component of an SNA network that can be assigned an address and receive or send information is known as a network addressable unit (NAU). The three major types of NAUs I'll discuss are logical units (LUs), physical units (PUs), and system service control points (SSCPs). Two NAUs must establish a *session*, or linkage, in order to be able to communicate on the network.

Logical units (LUs)

Nobody ever said that IBM terminology was easy or even that it made sense. The SNA world has its own terminology; it's important to first go over some key terms in order to consider how hosts can be linked to LANs. In the SNA environment, any device or application program that needs to use services is known as an *SNA user*. An SNA user accesses the network through a point of access known as a *logical unit (LU)*. Think of these LUs as logical ports and not physical connections. Table 6.1 lists the major LUs. We'll return to LU 6.2 later in this chapter when examining peer-to-peer relations.

TABLE 6.1 Major Logical Units (LUs) Under SNA

LU type	Description
Type 1	Supports communication between an application program and a remote terminal.
Type 2	Supports communication between an application program and a 3270 terminal.
Type 3	Supports Type 2 communication services for keyboard/printer terminals.
Type 4	Supports communication between host and terminal for IBM office equipment.
Type 6.1	Supports communication between applications running on hosts.
Type 6.2	Supports program-to-program communication without the need for a host.

TABLE 6.2 Key SNA Physical Units (PUs)

PU type	Description
Type 2.0	Cluster controllers to 3270 terminals
Type 2.1	IBM minicomputers and microcomputers (Type 2.1 nodes can communicate with each other with the host as an intermediary)
Type 4	Communications controllers functioning as front-end processors
Type 5	Host computers

Physical units (PUs)

The physical unit (PU) is the actual physical device or communication link found on an SNA network. The term *physical unit* actually refers to not just the physical device but also any software and microcode that defines the services performed by that device. Communications processors, cluster controllers, and terminals are all defined on an SNA network as PUs. Table 6.2 lists the key PUs.

System service control points (SSCPs)

The system service control point (SSCP) on an SNA network is the network addressable unit that provides the services necessary to manage a network or a portion of a network. The SSCP resides in the virtual telecommunications access method (VTAM) control program on the host computer. The SSCP activates and deactivates network resources under its control or "domain" with assistance from the network control program (NCP), a physical unit running on a communications controller (front-end processor). Before NAUs can communicate with each other, their physical and logical connections must be turned on by the SSCP, and the NCP must be loaded into the FEP. Figure 6.6 illustrates how these various NAUs work together to enable a 3278/9 terminal to access a program on the host computer.

SNA's layered architecture

SNA has its own layered architecture, which predates the OSI model, and, as shown in Table 6.3, there's no one-to-one layer correspondence between these two architectures. Despite their differences, SNA and OSI both enjoy the advantage of any layered architecture—the ability to modify or revise one layer's services without having to create an entirely new architecture. You've already seen how a gateway links LANs and mainframe computers; by examining the major functions performed in each SNA layer, it should become clear why micro/mainframe communications require a gateway that uses all network layers, rather than a router capable of functioning only at the network layer.

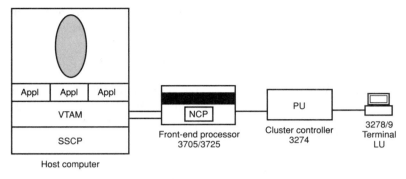

Figure 6.6 NAUs working together to enable a terminal to access a mainframe program.

TABLE 6.3 SNA's Layered Architecture

OSI model layer	SNA layer
Application	Transaction services
Presentation	Presentation services
Session	Data flow control
Transport	Transmission control
Network	Path control
Data link	Data link control
Physical	Physical control

The physical control layer corresponds the OSI model's physical layer. It provides specifications for both serial connections between nodes and high-speed parallel connections between host computers and front-end processors. The data link control layer corresponds to OSI's layer 2. The major protocol in this layer is IBM's synchronous data link control (SDLC), although there are also protocols for IBM's own Token-Ring Network as well as a high-speed parallel link protocol for IBM's S/370 hosts.

The path control layer handles routing and flow control. The protocols in this layer also handle sequencing and expedited network services. In the event that a specified route is not longer viable, this layer notifies the layer above it (transmission control). The transmission control layer contains protocols that provide the pacing for data exchanges between NAUs so information isn't lost. It also handles data encryption when this is specified. This layer adds a header to an SNA frame that includes its control information.

The data flow control layer corresponds roughly to the session layer of the OSI model. It's responsible for the establishment of half-duplex and full-duplex network sessions. The presentation services layer is responsible for formatting, translation, and other services associated with the way the data must look.

Finally, the transaction services layer corresponds to OSI's application layer. Here you'll find IBM's SNA distribution services (SNADS), which is a way of distributing data asynchronously between distributed application programs. IBM's document interchange architecture (DIA), also found in this layer, provides a way to ensure that documents can be exchanged among different application programs.

Systems Application Architecture (SAA)

Systems Application Architecture (SAA) is IBM's long-range plan and set of specifications to link together its entire computer family of products. SAA includes the common user access (CUA), common programming interface (CPI), and common communications services (CCS). If programmers follow the specifications for these three key interfaces, eventually it won't matter whether a user accesses a program on an IBM mainframe, minicomputer, or PS/2 running OS/2. All these programs will have the same "look and feel" and also be able to exchange information. Under pressure from large corporate customers with mixed computer environments, IBM has incorporated communications standards used by other vendors, as well as OSI model protocols, in SAA.

Many of IBM's largest customers are concerned whether or not SNA will ever be OSI-compatible; they fear being locked into IBM's proprietary network architecture. If a company were a 100% IBM shop, SNA would be vastly preferable to the OSI model's suite of protocols since it's optimized for IBM hardware and doesn't have the overhead of OSI protocols. In a mixed environment, however, companies are planning for when OSI's suite of protocols will serve as a common language, connecting a company's diverse computer resources.

IBM has reassured its major customers by including OSI protocols in SAA common communications support. It will offer the programming interfaces for the OSI communications subsystem as a feature of ACF/VTAM, the company's basic mainframe communications software. In fact, the OSI communications subsystem provides the base for the SAA OSI application layer functions. OSI file services implement the OSI FTAM protocol in the SAA environment. IBM has also indicated that SAA will support OSI protocols for LANs and WANs, CCITT X.400 electronic mail protocols, and CCITT X.500 protocols for a global e-mail directory.

NetWare and SAA mainframe connectivity

Novell designed its NetWare for SAA to support multiple host links and up to 1,000 simultaneous client sessions. The entire SNA protocol stack is incorporated into a network loadable module (NLM). The NetWare file server running this NLM can connect to an IBM host via a synchronous data link control (SDLC) link. It's also possible to create an indirect host-to-NetWare LAN connection by linking the file server to a token-ring LAN via a token-ring interface coupler (TIC).

Network managers can disable or enable host access to specific workstations, and limit users to certain types of LU sessions. Novell's communications services manager enables LAN managers to monitor the performance of SAA LAN-to-host sessions and track network statistics, including link efficiency and traffic patterns.

A Checklist for Gateways to IBM Mainframes

Now that you've had the opportunity to examine the basic components of an IBM mainframe system and its network architecture, and also survey the different types of gateways available, it's time to look at specific gateway features. This section will provide you with brief descriptions of the major gateway features you'll need to consider as a network manager or systems integrator in order to evaluate the wide range of gateways now available.

Number of simultaneous terminal sessions on a PC

Gateways provide a varying number of simultaneous (concurrent) terminal sessions on a single PC. Is a DOS session supported separately, or is it part of the total sessions permitted a terminal in the vendor's literature? The number of terminal sessions available to a PC range from 1 to 32 among the more popular gateway products. Is there a hotkey that makes it easy to switch back and forth from DOS to mainframe terminal emulation?

A gateway that emulates an IBM 3299 looks like eight terminals to a cluster controller. If each terminal can support five sessions, then the gateway provides a total of 40 sessions for distribution among network workstations. The total number of concurrent sessions supported by a gateway can vary widely, from less than sixteen at the low end to 254 at the high end.

Terminals supported

Some gateway software packages can emulate all members of the 3178/9 and 3278/9 families of terminals; others are limited to the more common models. An increasing number of LANs will require graphics terminal emulation in the near future because virtually all software is becoming more graphics-oriented. Does the software support IBM 3179G or 3279G graphics terminals?

File-transfer capabilities

While network managers have been able to transfer binary files back and forth between mainframes and micros using a tedious record-by-record approach, they've pushed for more efficient software that would let them encode, compress, and block data; calculate check sums; and then decompress, unblock, and decode the files at both ends. IBM's IND$FILE protocol now has become an industry standard. Some vendors, such as Data Interface Systems Corporation, provide their own proprietary file-transfer software as well as IND$FILE.

Some vendors improve the speed of IND$FILE by increasing the size of the PC gateway's buffer. Another key file-transfer feature for gateways is to permit file transfer to take place in a host session. This enables multiple file transfers to take place simultaneously at each workstation. These file transfers might be taking place between a workstation and different hosts or on the same host. The gateway's ability to perform file transfers in the background mode is important. If host-to-LAN file transfer is a major function for your proposed gateway, then it's important to consider the file-transfer protocols supported by the gateway.

Protocols supported

Virtually all gateways support IBM's synchronous data link control (SDLC) protocol. A number of gateways support binary synchronous control (BSC) protocol, an older half-duplex approach still found on a significant number of mainframes. Many remote sites still use remote job entry (RJE), a method of submitting work to a mainframe in batch format. Other key protocols you might need supported include X.25, VT100 for communication with DEC systems, and Burroughs for communication with older (pre-Unisys) Burroughs mainframes.

Support of programming interfaces

In enterprise networks of the near future, companies will want far more connectivity than simple terminal emulation. They'll want to automate log-on procedures, create custom screens, automate many of the LAN-to-host file-transfer tasks, and develop communication links between mainframe and LAN programs. Does your gateway support the necessary application programming interfaces (APIs) to enable your programmers to write the necessary code? IBM's high-level language applications programming interface (HLLAPI) enables mouse support.

A gateway's support for IBM's advanced program-to-program communications (APPC) protocol for distributed processing could also be important in the future. APPC is already being used on IBM's SNA distribution services (SNADS) to send documents and files back and forth between two different systems. IBM's distributed data management (DDM) uses APPC to transmit data between a client system and a database server. APPC will be a key component of SAA, and a tool for corporate programmers who want to establish links between programs running on different IBM systems.

APPC incorporates two relatively new SNA protocols. PU 2.1 permits two processors to communicate on a peer-to-peer basis, while LU 6.2 permits two application programs to have a peer-to-peer conversation. In the future, it's likely that programmers will use APPC and LU 6.2 to write code so two programs running on different machines, such as a host and a microcomputer, will be able to exchange information transparently to the computer user. LU 6.2 will also be used to write multiple front-end applications, including spreadsheets and databases, to run on client workstations that communicate via structured query language (SQL) commands with a shared database running on a server or host.

Dedicated and pooled LUs

Some gateways permit dedicated LUs, an approach that guarantees that a given LU will be available to a LAN node when requested. Pooled LUs are available to all LAN nodes on a first come, first served basis. On most gateways these pooled resources are freed-up and reusable once the user terminates a session.

Gateway management

The types of gateway management features supported vary widely among gateways. Some provide a gateway monitor, which enables a network manager to examine all session assignments and enable or disable devices. Some gateways let the network manager configure dedicated devices that will automatically attach to assigned devices when a workstation is initiated. Dynamic device attachment logic retains prior assignments for each device and then attempts to reserve a free device for its most recent user unless or until no other device is available for another attaching user.

LAN features supported

Since gateways were available before LANs became popular, some vendors have put more effort into making their products "LAN friendly" and other gateway vendors have provided minimal LAN functionality. Can the gateway recognize directory paths without needing to switch to the proper directory? Does the gateway communicate over NetBIOS or over Novell's IPX protocol? NetBIOS versions differ and problems could arise on a large NetWare LAN if the gateway doesn't support IPX. NetWare network managers also prefer gateway software that runs in a directory that can be set read-only. In other words, what has the gateway vendor done to make this product easier for network users to operate?

Remote speeds supported

If your gateway is to be remote, the transmission speed it supports is crucial. Some gateways still support only 19.2-Kbps modems, while others support up to 64-Kbps transmission speeds. You might have to consider multiple remote gateways to reduce traffic congestion.

Summary

There are a number of different types of LAN gateways to IBM hosts, including local DFT coaxial connection, direct connection via token-ring interface coupler (TIC), remote link via modem, and X.25 connection. Mainframes can also be used as routers to link two LANs that have mainframe gateways.

In order to understand mainframe gateways, it's necessary to know something about IBM's systems network architecture (SNA), its layered network architecture. Historically this has been a master/slave system with all communications initiated by the host. IBM's LU 6.2 and APPC make peer-to-peer communications possible in the future between LAN programs and mainframe programs.

IBM's systems application architecture (SAA) is a set of specifications for consistent user and program interfaces for its entire computer family. The release of SAA-compliant services on NetWare will facilitate this communication between LANs and mainframes.

7

Wide Area Networks

In this chapter, you'll examine:

- Metropolitan area networks (MANs)
- Wide area networks (WANs)
- T-1 links
- Packet switched networks including frame relay
- Switched multimegabit digital service (SMDS)
- SONET

Networks are no longer limited to one site; they can be city-wide (a metropolitan area network) or global (wide area networks). In this chapter, I'll describe the nuts and bolts associated with building large networks, and a number of different transmission options ranging from T-1 links to packet-switched networks. Then I'll show you some evolving technologies that could dominate the next decade. This chapter is a survey. Reading it won't make you an expert on wide area networks, but it should help you understand some key concepts and learn some of the jargon associated with this complex topic.

Metropolitan Area Networks (MANs)

The first chapter described several different types of local area networks, workstations linked together at a single site. Later in this chapter you'll examine wide area networks, networks covering hundreds and even thousands of miles. In between LAN and WAN lies the *metropolitan area network (MAN)*, a network that covers an entire city.

The IEEE 802.6 committee has developed recommendations for a MAN that incorporate the concept of dual counter-rotating rings or buses similar to FDDI. MANs are designed to act as digital backbones that link together LANs throughout the city. They're designed to handle voice, video, and data traffic at speeds in excess of 100 Mbps and can incorporate fiber-optic and coaxial cable, and even radio transmission as media.

A MAN is composed of dozens of subnetworks that communicate with each other through bridges, routers, and gateways. The "glue" that makes all this work is a protocol known as distributed queue dual bus (DQDB). As Figure 7.1 illustrates, DQDB consists of a dual bus topology with traffic traveling in opposite directions. Fixed-length "slots" originate at the head of a bus and terminate at the end of a bus. An access unit (AU) attaches a network workstation to the dual bus cabling. The AU contains the protocol necessary to perform DQDB functions. Malfunctioning AUs are ignored by the network.

Nodes use a QA slot, a packet that includes a header and a data field. Two key fields in this packet include a busy bit, which indicates whether or not a slot is empty, and a request field, which indicates whether the slot is being queued for transmission.

Nodes set bits on QA slots traveling on one bus to reserve a slot traveling on the other bus. Nodes maintain a counter that tells them how many requests are pending in a queue; hence, they know how many empty slots a node must ignore before it becomes its turn to access the slot. The advantage of this type of approach is that the slot is constantly in use; there are always nodes with reservations waiting their turn. To get more information on available MANs in your area, talk to your local phone company. Pacific

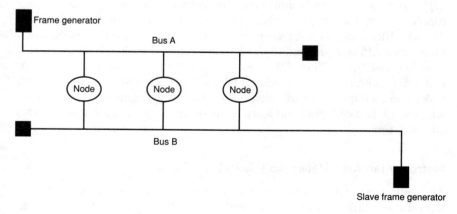

Figure 7.1   The IEEE 802.6 dual bus topology.

Bell, Atlantic Bell, etc., are promoting these city-wide networks using their enormous switching capacities.

Proteon and the Australian company QPSX have developed a multiprotocol router to function as an interface between 802.6 systems and FDDI. This announcement is significant because some large companies are likely to adopt FDDI as their standard for corporate network backbones. The link between FDDI and 802.6 will help to link LANs, MANs, and WANs.

Recent conventions have illustrated how far MANs have come. AT&T and four regional Bell operating companies have demonstrated switched multimegabit data service (SMDS), which is the best known of current MAN services, and Siemens has demonstrated its QPSX MAN switch. Significantly, one of Siemens' demonstrations featured an Ethernet LAN connected to a MAN switch. This switch was connected to another MAN switch, which, in turn, was linked to a second Ethernet LAN. Another Siemens' demo featured a Macintosh microcomputer linked to a remote mainframe computer via a MAN. The two computers ran graphics-oriented programs in a client-server environment. These demonstrations illustrated why interest in MANs is growing; they have the speed and bandwidth to link together LANs as well as incorporate host computers in city-wide networks.

Wide Area Networks (WANs)

Wide area networks link together networks located in different geographic areas. A company might have to link together its Ethernet LAN in Boston and its token-ring network in Los Angeles. Another company might want to link together several sites so they can exchange both voice and data information.

One major problem for network designers is that, while PC-based local area networks and mainframe networks both routinely transmit data at several million bits per second, transmission speeds over phone lines have lagged considerably. Linking two 10-Mbps Ethernet LANs with a 19.2-Kbps analog phone line is bound to create a serious traffic bottleneck. Many companies still struggle with 9,600-bps modems. This congestion problem is becoming more manageable because of the growth of digital transmission services, as well as the development of more sophisticated types of services, such as frame relay, which I'll discuss later in the chapter.

A major problem for network managers and network designers is that WAN links are inherently less reliable than local transmission links. Error checking might not be crucial on a token-ring LAN, but it becomes a major concern when data is transmitted over phone lines that are subject to all kinds of electrical interference. When remote communications are disrupted and data has to be retransmitted, the LAN-to-WAN link becomes even more congested and inefficient.

The effect of divestiture on wide area networks

The breakup of the Bell system dramatically changed the telecommunications industry and, in turn, how companies transmit their voice and data on wide area networks. Prior to divestiture, companies dealt exclusively with the Bell system. Today, network managers as well as systems integrators designing wide area networks must deal with several different vendors. From the customer's premises to the closest central office (CO) is the province of the local phone company, a *local exchange company (LEC)*. LECs include the Bell operating companies (Pacific Bell, Atlantic Bell, Southwest Bell, etc.) as well as a number of independent companies.

Here's where the telecommunications jargon becomes a bit overwhelming, but it's important to understand who you must deal with in order to send voice or data from one network to another network. The LECs control all calls made within their geographic areas or *local area transport area* (LATA). An LEC provides service within all the LATAs within its territory, but it cannot route a call from one to another without going through an interexchange carrier's network. Within each LATA are interface points to the inter-LATA carriers, known as a *point of presences (POPs)*. Each inter-LATA carrier, such as AT&T, MCI, or Sprint, have their own lists of POPs. AT&T calls its POP a *central office (CO)*. These POPs are the only places within a LATA where an inter-LATA carrier can receive and deliver traffic. Figure 7.2 illustrates the route a call takes from one LATA to another.

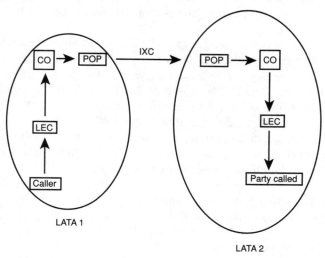

Figure 7.2 The route a call takes.

Digital signal transmission

In 1957 the Bell system installed its first T-1 trunk to carry high-speed digital voice signals. Since we're talking about digital and not analog signals, a device is needed to generate these digital signals. The digital pulses generated at a customer's site have to be filtered effectively to eliminate noise and distortion. Customers interface with the telephone company's digital network through a channel service unit (CSU) or a data service unit (DSU).

Both the CSU and the DSU function as digital modems, though the CSU also provides some line conditioning and diagnostic functions. A T-1 trunk contains 24 channels, and each channel is capable of handling 64,000 bits per second. An additional 64,000 bps is required for error checking, so one T-1 line requires a bandwidth of 1.544 Mbps:

$$64,000 \text{ bps per channel} \times 24 \text{ channels} = 1.536 \text{ Mbps}$$

$$8 \text{ bps per sample} \times 8,000 \text{ samples per second} = 64,000 \text{ bps}$$

$$\text{Total} = 1.544 \text{ Mbps}$$

This 1.544-Mbps rate is known in the telecommunications industry as DS-1 (digital signal, level 1). There's an entire hierarchy of digital signal bandwidth options available:

Signal	Speed	Number of T-1 channels
DS-1	1.544 Mbps	1
DS-2	6.312 Mbps	4
DS-3	45 Mbps	28
DS-4	274 Mbps	168

The T-1 (DS-1) and T-3 (DS-3) circuits are popular. The European digital hierarchy is slightly different from this North American standard. In Europe DS-1 consists of 32 channels, each of which transmits at 64 Kbps, for a total bandwidth of 2.048 Mbps. Two of these channels are used only for signaling and network management functions.

Why T-1 Circuits Are Popular

There are a number of reasons why T-1 (and by extension T-3) circuits are so popular. Digital transmission produces much higher quality voice signals than analog transmission. As mentioned earlier, companies can save substantially by consolidating their voice and data transmissions over the same circuit rather than maintaining two separate transmission paths.

Many T-1 multiplexers provide redundancy by offering automatic alternate routing so other circuits are used if a path is out of service. Another advantage of T-1 service is its flexibility. The bandwidth can be allocated in different ways depending on voice and data needs. For example, a typical multiplexer might permit the user to program data channels from 300 bps to 1.5 Mbps. Voice channels can be programmed to transmit at 64 Kbps or at 24 or 32 Kbps using different compression schemes.

T-1 multiplexers usually have both synchronous and asynchronous data interfaces. A network manager with several low-speed data terminals could take advantage of a multiplexer to consolidate these low-speed data sources into a single DS-0 channel. Figure 7.3 illustrates how a subrate multiplexer can fill a single DS-0 channel efficiently. Figure 7.4 illustrates how a subrate multiplexer can send information from 12 different data circuits to a T-1 multiplexer. This data stream then goes through a digital access and cross-connect service (DACS), which routes it to the appropriate T-1 multiplexers to complete the journey.

Another nice feature of T-1 service is that you can expand lines easily by adding a circuit card to the T-1 multiplexers at each end of transmission. A small multiplexer can support up to 16 lines, while larger multiplexers can support up to 128 lines.

T-1 frames

A T-1 trunk carries a serial bit stream that's transmitted using time-division multiplexing on a frame-by-frame basis. A frame consists of 192 bits (8 bits × 24 channels) plus one additional synchronization bit, for a total of 193 bits.

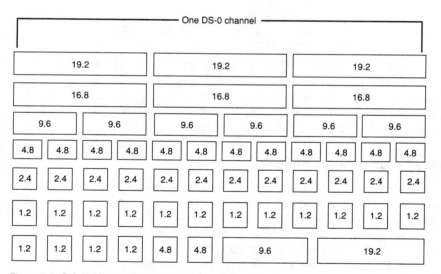

Figure 7.3 Subdividing a single DS-0 channel (*CoastCom*).

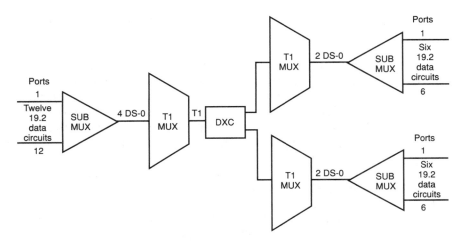

Figure 7.4 Consolidating several low-speed data streams with a subrate multiplexer (*Coast-Com*).

Before divestiture, AT&T used T-1 lines for internal purposes, but customers were offered voice-grade private lines (VGPL) for voice transmission, and dataphone digital service (DDS) for data transmission. When AT&T tariffed T-1 service in 1983, it became possible for corporations to combine their voice and data traffic over T-1 lines and save a considerable amount of money.

Over 30 vendors offer T-1 equipment today. In order to understand the specifications for these different products, it's necessary to understand how the data stream is framed, how error checking is performed, and the type of compression scheme (if any) used to transmit the data.

The standard D-4 super frame

The telecommunications standard D-4 "super frame" framing scheme consists of creating a super frame of 12 separate 193-bit frames. As Figure 7.5 illustrates, a framing bit identifies both the channel and the signaling frame.

The extended super frame

For the past several years AT&T has advocated a new framing scheme known as the extended super frame (ESF). As illustrated in Figure 7.6, this type of framing consists of 24 frames grouped together with an 8,000-bps FE channel used as follows:

- 2,000 bps for the framing and signaling found under D-4
- 2,000 bps for CRC-6 error checking

- 4,000 bps as a data link to provide additional diagnostics and network management functions

The 4,000-bps data link can handle the accumulation of performance data as information on logic errors. This information is stored in registers that can be accessed by the carrier or by the customer. Figure 7.7 illustrates how this network management information is accessed. The channel service unit (CSU) pictured here serves as the interface between the customer's terminal equipment and the carrier. Installed at the end of each T-1 circuit, the CSU manages and monitors each of the T-1 channels. It also handles ESF framing, including the CRC error checking function and the data link information. AT&T currently uses ESF to monitor and maintain its T-1 lines without charge to users.

ESF is growing in popularity because the diagnostics it provides results in less down-time. In addition, the CRC error rate indicates any degradation in transmission performance, a sure sign that repairs need to be done before conditions become worse.

Time-division multiplexing: the key to T-1 transmission

Data from a number of different sources feeds into each of the 24 DS-0 channels comprising a DS-1 line. This data must be placed in the appropriate super frame or extended super frame format, and then transmitted via a T-1 multiplexer using time-division multiplexing (TDM). Time-division multiplexing consolidates the data stream from each of the DS-0 channels by using an approach that guarantees a time slot for data from each of these channels. As Figure 7.8 illustrates, it's possible for a channel to not have any-

Framing bit

| 1 | F1 | 0 | F2 | 0 | F3 | 0 | F4 | 0 | F5 | 1 | F6 | 0 | F7 | 1 | F8 | 1 | F9 | 0 | F10 | 0 | F11 | 0 | F12 |

Each frame contains 193 bits

Figure 7.5 A standard D4 super frame.

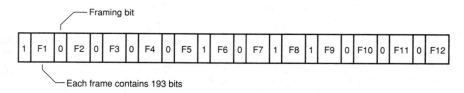

Extended super frame = 24 frames.
Sync bit pattern at head of frames 4, 8, 12, 16, 20, and 24.
Six CRC bits at head of frames 2, 6, 10, 14, 18, and 22.
Twelve bits in odd = numbered frames (DL) not currently used.

Figure 7.6 The ESF framing pattern.

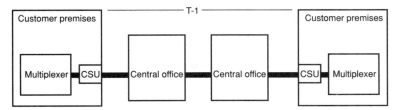

Figure 7.7 End-to-end performance monitoring with ESF.

Channels

1		Asynchronous data			
2	Synchronous data	Synchronous data			
3		Fax		Fax	
4	Voice	Voice			
5	Voice	Voice	Voice	Voice	
6	Asynchronous data			Asynchronous data	
7	Synchronous data	Synchronous data			
8	LAN data	LAN data		LAN data	

Figure 7.8 Time-division multiplexing.

thing to send when its turn arrives; under such circumstances the time slot remains empty. When you look at fast-packet T-1 technology later in this chapter, you'll see how this new approach overcomes this TDM drawback.

Remote bridges and wide area networks

Time-division multiplexing is often used to build a wide area network. Sometimes a WAN consists of several LANs remotely bridged together. The major advantage of remote bridges that link together LANs located at different sites is that they can function as a single, seamless wide area network independent of all upper-layer network protocols, including TCP/IP, DECnet, SPX/IPX, and Vines IP. A well-designed WAN might require the integration of different higher-level protocols, different transmission speeds, and different transmission media.

Figure 7.9 illustrates a wide area network linking together several different Ethernet networks. Notice that two networks are linked by two 64-Kbps lines, which results in load sharing. Two networks are joined together by 100-Mbps fiber-optic connections and another WAN link consists of a T-1 line.

Digital access and cross-connect systems (DACS)

You've been looking at T-1 transmission in terms of T-1 multiplexers, but there are situations where a digital access and cross-connect system (DACS)

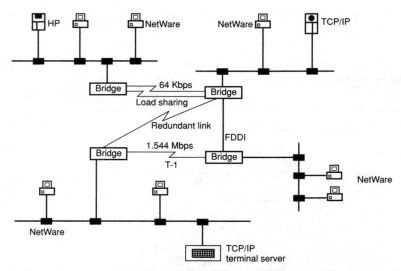

Figure 7.9 Remote bridges using a wide range of links and transmission speeds to form a WAN.

can serve the purpose at a much lower price. The telephone company first used DACS in its central offices to switch DS-0 channels from one T-1 span (DS-1) to another. DACS were designed to break down a DS-1 stream into 24 DS-0s. These DS-0 channels were then routed to time slots on any other DS-1.

The limitations of a DACS include its inability to deal with a subchannel of less than 64 Kbps, a topic we'll examine when looking at fractional T-1 service. Another DACS limitation is its lack of sophistication when it comes to automatic alternate routing. Traditional DACS require that this routing be done manually at each switch. Finally, DACS lack the ability to compress voice transmission; this means that a network manager can't make efficient use of DS-0 channels by using subrate voice channels at 16 or 32 Kbps.

Despite these limitations, a DACS serves a very important function. A DACS in conjunction with a T-1 multiplexer can link IEEE 802.3 LANs to form an Ethernet T-1 WAN. A LAN via an Ethernet transceiver connects with a DACS via an RS-449, V.35, or X.21 interface on the other side. Let's look at a practical application of such a device. Assume offices in four different cities need to be linked. Functioning at the bridge level, this WAN is completely transparent. It operates independently of all higher-level protocols, such as XNS, DECnet, and TCP/IP. Note that it's a special multiplexing device that takes the wide range of data streams from the four different company branches in four different cities and places them in the traditional DS-1 format required for the DACS.

Selecting a T-1 multiplexer

T-1 multiplexers have gone through several generations; in fact, they seem to evolve almost as quickly as fruit flies. First-generation multiplexers were known as *channel banks*. Designed strictly for super frame or extended super frame formats, these devices divided the T-1 bandwidth into 24 64-Kbps channels and didn't permit any adjustment in the size of the channels.

Second-generation T-1 multiplexers added the ability to send voice information at lower rates. These schemes for adjusting the speed of voice transmission (16 Kbps, 32 Kbps, etc.) are generally proprietary, so it isn't practical to try to mix and match these multiplexers. Another feature of second-generation multiplexers is their ability to adjust the size of a channel according to the manufacturer's range of options. A network manager with several low-speed transmissions from terminals, for example, could send these transmissions over a T-1 line.

Third-generation T-1 multiplexers include the features found in the previous generation, but add some features that make them ideally suited for large networks. One feature is the number of T-1 aggregates the multiplexer can support. An aggregate is composed of the composite data from all the channels comprising a T-1 link. Since network operations often require several data streams and multiplexers to transmit streams to a centralized multiplexer, this device must be able to handle such a load and keep track of all the corresponding channels.

Often data enters a T-1 multiplexer as a channel on one aggregate and must be routed to another aggregate without being absorbed by the channel side of the multiplexer. In other words, it's simply passing through the multiplexer and being rerouted to another multiplexer. This type of transmission is often referred to as "drop and insert." T-1 multiplexers vary in the delay time required for this rerouting.

Closely related to this delay time is the way the T-1 multiplexer routes frames. With some multiplexers, the routing information is maintained in a central node, while in other multiplexers it's distributed among nodes. The distributed approach takes longer because information must be exchanged across the network before routing can take place. The trade-off is that, when all routing information is maintained in a central node, the failure of the node means the failure of all routing capability for the multiplexer and network.

Another feature distinguishing third-generation T-1 multiplexers is network management. These devices tend to support a number of network management protocols and provide graphics-oriented displays of an entire wide area network rather than the activities of the single multiplexer. This approach enables a network manager to see at a glance where a problem is located and take appropriate action.

Another significant feature to look for in a T-1 multiplexer is its support of protocols and standards. While super frame and extended super frame formatting is essential, the appropriate interface for integrated services digital network (ISDN) could be important in the not too distant future. We'll examine ISDN in chapter 8.

The type of routing performed by the T-1 multiplexer is also important. While many devices offer dynamic routing, some use a static routing table that the network manager must create and maintain. On a large network, this approach could prove to be inefficient and unresponsive to changing conditions.

Closely related to changing conditions on T-1 lines is the ability of a T-1 multiplexer to dynamically reroute information when a line fails. Time-sensitive applications such as those found under IBM's systems network architecture (SNA) cannot maintain a session during the period of time normally required for a manual reconfiguration of a T-1 network. Lengthy rerouting of voice channels can also have a negative effect. Many industry experts use five seconds as the maximum amount of time a caller will wait before hanging up.

While some T-1 multiplexers are still designed to serve only as point-to-point devices, more are now full-fledged network models.

Fractional T-1

A recent innovation in T-1 transmission has been fractional T-1 service (FT1). Common carriers subdivide T-1 bandwidth into its 24 DS0 channels of 64 Kbps. The user can bundle together these channels to fit particular applications. A customer might require one eighth (192 Kbps), one fourth (384 Kbps), or one half (768 Kbps) of the available bandwidth. Users pay for only the bandwidth they need.

Voice transmission can be compressed a good deal without losing any content (think of all the "ahhs" and "uhhs" in a typical conversation). With popular compression schemes, it's common for customers to take a 384-Kbps FT1 channel and use it for 20 voice lines of 16 Kbps each, or perhaps 10 voice lines of 32 Kbps apiece. Because of the efficiency of these compression schemes, no meaningful content of a conversation is lost.

While large corporations can realize tens of thousands of dollars a month in savings using FT1, they could save even more money if local phone companies would cooperate. Presently, customers must pay for a full T-1 circuit from their premises to the interexchange carrier's point of presence (POP), even if the long-haul portion of the circuit (POP to POP) is fractional. Figure 7.10 illustrates how fractional T-1s can be used. The following are some major fractional T-1 domestic and international offerings:

- International record carriers (RCA, Wordcom, and IDB)
- AT&T International Accunet
- AT&T International Skynet

- Sprint meeting service
- A spectrum of digital services
- Cable and wireless fractional service
- Williams Telecommunications fractional service
- MCI fractional service
- Lightnet
- Teleport
- RBOCs

Imagine that you have nine sites, each of which has a LAN, that need to be linked together in a wide area network. The higher-level protocols at the various sites include NetWare's IPX, 3Com's XNS, and TCP/IP. The MAC layer of all these networks is IEEE 802.3. The corporation has looked at AT&T's 56-Kbps dataphone digital service transmission, but has rejected it as far too expensive. The company has also considered V.35 interface cards and concluded that the price is prohibitive.

One possible solution is a new type of WAN multiplexer that combines the best features of a learning bridge with T-1 and fractional T-1 service. This bridge enables a network manager to run multiples of 64-Kbps circuits, up to the maximum T-1 capacity of 1.544 Mbps. In the case of the corporation with corporate headquarters that need to be connected to nine sites, The bridge could allocate nine DS-0 bundles of 64 Kbps each to establish the nine separate 64-Kbps bandwidth links. The beauty of this type of wide area network is that, since it's essentially a remote bridge using fractional T-1 service, the workstations at any given site view the entire WAN as one large, transparent network.

T-3 multiplexers

Some companies use T-3 multiplexers to consolidate T-1 and fractional T-1 networks. It's definitely economical to use T-3 circuits rather than multiple T-1 circuits. If multiple T-1 lines are going to the same destination within a pathway, they can be combined and transmitted over a T-3 circuit. For relatively short distances, many industry experts use a 5-to-6 cost ratio, and

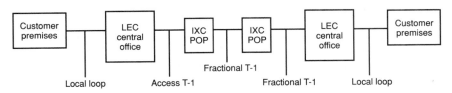

Figure 7.10 Use of fractional T-1.

an 8-to-10 cost ratio for longer distances. This means that, for relatively short distances, it becomes cost-effective to use a T-3 line to replace 5-to-6 T-1 circuits. The added bandwidth is virtually free and can be used to carry additional voice traffic. Some companies that have traditionally used separate lines for voice and data have consolidated these two different streams using a T-3 line.

Another use of T-3 is as a hub that takes a T-1 channel and switches it to another T-1 channel. Such devices provide key network management features that make it easier to manage a wide area network. A network manager can monitor and control any T-1 channel within a T-3 circuit. Test devices internal to the hub provide remote testing and monitoring.

The IEEE 802.1G committee and remote bridge standards for WANs

The IEEE 802.1G committee has been working to develop a standard for remote bridges. Perhaps because of conflicting interests by members who represent competing vendors, the committee has opted for the remote bridges to use the same spanning tree protocol adopted as a standard for local bridges.

This approach doesn't work very well for wide area networks, however. The spanning tree protocol selects one path and places all other routes in standby mode. Unfortunately, on wide area networks it's wasteful and not efficient to have several links sitting idle.

Different vendors have solved this problem by letting these standby-mode links provide load sharing across the WAN as well as additional system fault tolerance. Unfortunately, each vendor has chosen its own proprietary protocols to achieve this effect without eliminating spanning tree on the LAN side of the LAN-WAN connection. The result is a lack of interoperability on the WAN side of the LAN-WAN bridge, and a difficulty in mixing and matching remote bridges from different vendors.

Network designers who see more of a need in the future for WAN connections might opt for routers today because routers have no problems with alternate paths. They might be slower and more expensive, and have the additional limitation of being protocol-specific, but they won't lock a company into a vendor's proprietary technology and leave the company in a dead-end if and when a standard evolves for remote bridging.

Packet-Switched (X.25) Networks

One of the major advantages of international standards in data communications protocols is illustrated by public data networks. These networks switch data in the form of CCITT X.25 packets across the country and around the world. The CCITT X.25 recommendation is based on the first three layers of the OSI model.

Typically, a DTE is connected to a packet assembler/disassembler (PAD), which provides the protocol translation from a data stream's native protocol (SNA, asynchronous, etc.) to X.25 protocol. At the destination end of the transmission, the PAD translates the packets from X.25 protocol back to whatever protocol is required. PADs help make X.25 transmission economical by concentrating the data streams from several DTEs.

The PAD must place the data it receives in a packet that contains control information for error checking and sequencing. Packets have a destination address; at each stage of their transmission route, they're checked for errors before being forwarded using the best available route at that particular moment. If a node receiving a packet detects an error, it requests that the sending node retransmit it. The path an individual packet takes is determined by the switching equipment, and no two packets will take the same route. As Figure 7.11 illustrates, it's not just likely but probable that packets will arrive at their destination out of sequence and need to be placed back in their proper sequence. All of this activity is transparent to users. Notice that packet-switched networks are usually depicted as clouds.

One key consideration in considering packet-switched networks is that the cost of using such a network is based on the amount of information transmitted and not on the distance between locations. Cost is usually based on the number of packets sent.

X.25 networks are also appealing for companies that have geographically dispersed heterogeneous networks. A company might have an SNA network at one location, for example, and a pre-SNA network at another

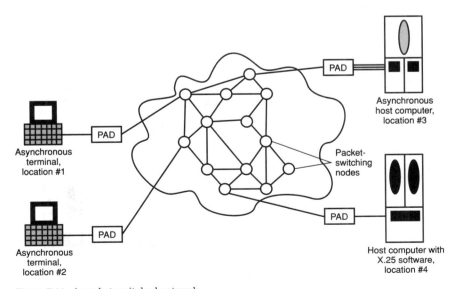

Figure 7.11   A packet-switched network.

site comprised of pre-SNA bisynchronous 3270s, 2780/3780 remote job entry devices, and asynchronous terminals. A third site might have DEC computers running VMS. The protocol conversion required for communications among these disparate networks can be handled by an X.25 network vendor.

The X.25 standard

The CCITT X.25 standard specifies the guidelines for computers and terminals to communicate with packet-switched nodes. It defines how non-packet equipment (DTEs) can connect to packet-switched networks through the use of a packet-switched node (DCE). Since this is an international standard, X.25 networks are used by many large corporations that have global communications needs. A U.S. packet-switched network can exchange packets with a network in South Africa, England, or Japan using a CCITT X.75 recommendation that defines what's required for a gateway between X.25 networks. Figure 7.12 illustrates how a company would transmit data from a host computer at a California site to a computer located in Connecticut.

Packet-switched network vendors and services

There are a number of major players in the U.S. packet-switched network market, including AT&T and the regional Bell telephone companies. Most vendors support transmission speeds ranging from 110 bps to 64 Kbps. Most packet-switched data network vendors also support a number of different protocols and offer protocol conversion, including 3270 bisynchronous, SDLC, 2780, 3780, and HASP.

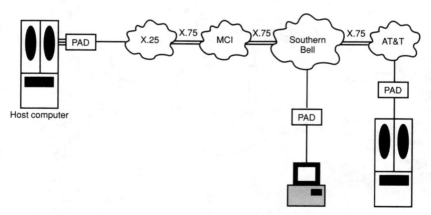

Figure 7.12   X.25 networks connected by X.75 gateways.

Selecting a public packet-switched network vendor

Public X.25 network vendors offer a number of features that might help distinguish one from another. This section lists some of the features to consider when selecting a public packet-switched network.

Throughput. Just as bridge and router vendors fudge a bit with their throughput figures, packet-switched network vendors also have to be questioned closely. Do their throughput figures count only data packets or do they include control information as well? How large are the packets? See if you can get a figure based on 128-byte packets since this is the industry norm.

Protocols supported. I touched on this topic earlier in the chapter. You or your client will know what protocols need to be supported; what vendors offer varies widely. While virtually all networks support IBM's SNA/SDLC and 3270 BSC protocols, only a few support Tandem, Uniscope, and X.400.

Maximum speeds supported. Most vendors offer dial-up lines ranging from 2,400 to 9,600 Kbps, along with dedicated lines ranging from 19.2 to 64 Kbps. A few vendors offer a 1.544-Mbps transmission speed for dedicated lines.

Error correcting. All vendors offer some kind of error correcting. Some provide special ports on both dial-in and private port bases that are equipped with error-correcting modems. The most popular error-correcting modems found in this environment are usually X.PC and MNP.

Network management. Will the vendor provide you with tools for network management and let you decide what paths packets will take based on the time of day? Will you be able to perform diagnostics from your own site and select alternate routing based on trunk speed?

Private packet-switched networks

Some companies might need a packet-switched network in an area where there are no public packet-switched services. It's also possible that they might have so much traffic between two sites that they could change substantial tariff charges by building their own private network. A third advantage to a private packet-switched network is that it provides centralized management and control, as well as increased security since the company has total control over all traffic on the network.

Hughes Network Systems is a major vendor of private packet-switched networks. Its integrated packet network (IPN) is used by a number of major corporations, including Hewlett-Packard and Ford Motor Company. The Hughes packet-switched network supports both earth and satellite transmission links. To provide the centralized management and control functions that characterize private networks, IPN uses a network control system (NCS) that manages distributed databases and coordinates administrative functions. NCS computers are designed to operate independently, yet remain fully linked for load-sharing and backup purposes. IPN offers enhanced security features, such as multilevel security that restricts user access and onboard encryption.

Northern Telecom is another major player in the packet-switched network industry. Its DPN-100 includes a software package called DPN Lanscope, which is designed specifically to enable users to manage LANs connected over a wide area network. DPN Lanscope provides fault and performance monitoring, resource management, software distribution, and usage tracking for geographically dispersed LANs. The software is used in conjunction with DPN Advisor, which resides on a UNIX workstation and provides graphic displays of the entire WAN for real-time control and fault management.

Hybrid Networks

For many companies, hybrid networks represent the best of both worlds. Since public packet-switched networks are traffic-sensitive, some companies use these public services where they're economically beneficial to handle light traffic between sites. Then they use X.75 gateways to link the networks to their own private networks where traffic is very heavy. Figure 7.13 illustrates a typical hybrid network.

X.25 Packet Switching versus T-1 Multiplexing

It can be a bit confusing differentiating between the two major methods of transmitting data on a wide area network. You've already looked at the T-1 multiplexer and the basic technique it uses, time-division multiplexing. Each channel on a time-division multiplexer is allocated a portion of the bandwidth, with that portion of the bandwidth totally dedicated to that particular channel. T-1 lines are cost-effective only up to a certain distance. Since a company pays for the use of a line 24 hours a day, the higher the usage, the more cost-effective the transmission link becomes.

Packet switching uses a technique known as *statistical multiplexing*, which means that bandwidth isn't permanently allocated to any given channel; instead, it's dynamically allocated using statistical algorithms to each channel based on that channel's need at any given time.

This ability to dynamically allocate bandwidth is particularly efficient when dealing with "bursty" traffic. A fax transmission, for example, might require massive amounts of bandwidth, but then that channel might remain idle for a substantial period of time.

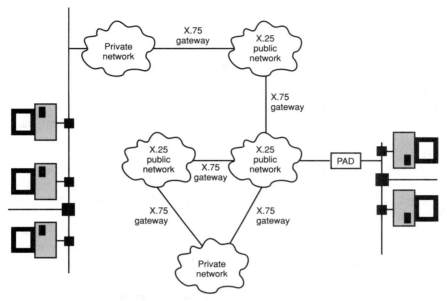

Figure 7.13 A hybrid packet-switched network.

Every advantage has its price, however. In order to be able to dynamically allocate bandwidth, a packet-switching statistical multiplexer must maintain constant communication with each channel to monitor what that channel's needs are; this communication requires overhead and causes some delay time.

If data flow is reasonably constant on different channels, then time-division multiplexing has the edge. T-1 multiplexers were designed to handle voice traffic, and it should come as no surprise that they excel in this area. The various compression schemes used with voice traffic still require a 64-Kbps channel; they simply provide more voice conversations over this channel.

A T-1 multiplexer requires less overhead and can be considerably faster. Unlike the T-1 speed of 1.544 Mbps, packet-switched networks using statistical multiplexing generally operate at a maximum speed of 64 Kbps. T-1 multiplexers aren't sensitive to protocols because they simply plug their control information into a preassigned slot and don't worry about what protocol the data uses.

What packet-switched networks do well is dynamically allocate their resources based on need and then route the packets efficiently. If a virtual circuit is tied up, they simply route packets along a different circuit. Since public packet-switched networks generally charge by the packet, packet switching can be much more cost-effective for transmitting small amounts of data. Also, the many services offered by packet-switched networking companies, including protocol conversion, encryption, etc., can help a company overcome incompatibilities between sites while maintaining security.

Creating a WAN by Bridging LANS Via X.25

Imagine a company that has several sites scattered across the country. These sites have Ethernet LANs, but the network operating systems vary and include an IBM LAN server, Novell's Netware, Apple's AppleShare, and Banyan Vines. The sites are scattered throughout the country, so the cost of T-1 links would be prohibitive.

As Figure 7.14 illustrates, the sites could be linked together using X.25 Ethernet bridges; since the LANs all use Ethernet, using X.25 remote bridges eliminates any concern for the different higher-level protocols running on these networks.

Let's consider a different scenario for a WAN. Instead of all the LANs having the same media access, let's assume a variety of different access methods but the same upper-level protocol, in this case NetWare and IPX. Say

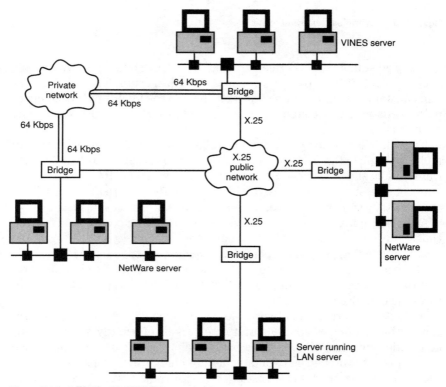

Figure 7.14 A WAN with X.25 Ethernet bridges.

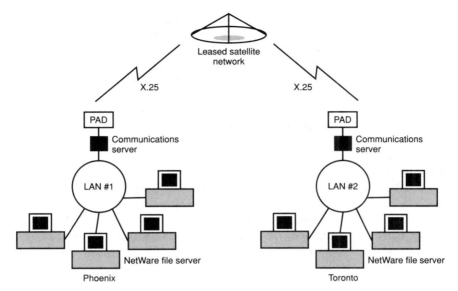

Figure 7.15 A WAN incorporating LANs and satellite transmission.

the company has its corporate headquarters in Dallas, regional branches at 35 sites, including New Orleans, San Francisco, and Chicago, and Canadian headquarters in Montreal. Each site has its own NetWare LAN. These LANs use whatever media access method is cost-effective at that particular site, including Ethernet, Arcnet, and token-ring networks. The company needs to link all sites together for electronic mail, LAN maintenance, and updating key information, including scheduling. A major problem for this company is that many of the sites don't have local public packet-switched services available.

As Figure 7.15 illustrates, this particular company can combine a number of different technologies to build its WAN. Each site's NetWare LAN uses a communications server with an X.25 gateway card that can transmit at up to 64 Kbps. Each site uses a multiplexer to transmit the packets to a modem, which in turn transmits the signals to an RF (radio frequency) format that's transmitted to a satellite dish, which uplinks the information to a leased satellite. The satellite serves as a backbone connecting all sites.

Fast-Packet Technology

Fast-packet technology is a relatively new development in T-1 transmission that combines the best features of T-1 multiplexing with some of the advantages of packet switching. Special T-1 multiplexers generate fast packets destined for a single channel only. The multiplexer allocates bandwidth instantaneously based on the data streams it receives. Let's assume that a multiplexer must handle voice, LAN traffic, and fax transmissions. This type of data mix would really benefit from a multiplexer that could allocate a single very large channel for a short period of time to expedite the transfer of a large amount of urgent data.

Besides the ability to dynamically allocate resources based on need, another reason why fast-packet technology transmits packets that are truly "fast" is that, when dealing with voice transmissions, this type of multiplexer frames voice bits only and filters out the silence. It then compresses what's left and repackages the bits into packets. Each packet contains an address up front. Notice that fixed time slots and fixed channel allotments aren't required because the fast packets have addresses indicating where they're destined.

Another major advantage of fast-packet technology is that the multiplexer can be programmed to recognize which data streams are crucial and send them first; voice information, for example, might be given higher priority than a terminal's transmission. Figure 7.16 illustrates fast-packet technology in action.

While X.25 packet-switched networks use layers 1, 2, and 3 of the OSI model, fast-packet technology uses only the first layer and a portion of the second layer. Since the frames contain their own control information for source and destination addressing and error detection, the fast-packet switches need to look at only the destination addresses and then pass the frames along quickly. Rather than include requests for retransmission of data with errors, the destination node simply discards data that contains errors. Less processing results in increased speed. While X.25 has a maximum transmission speed of 64 Kbps, fast-packet technology can provide a rate of up to 2 Mbps.

Frame Relay Technology

A new standard has recently evolved for using fast-packet technology more efficiently. Frame relay is a data link layer protocol that defines how vari-

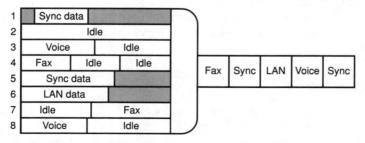

Figure 7.16 Fast-packet technology.

able-length data frames can be assembled. Frame relay requires only 48 bits of overhead, four to five times fewer bits than required with packet switching. Frames can be sized appropriately for the data loads they need to carry. The frame relay standard provides for a data link connection identifier (DLCI), which serves as an address field to allow for multiple logical sessions per physical data link. A standard frame format is specified for each of the various packet subsystems that will receive packets. Frames are labeled with the appropriate DLCI so, when the fast-packet system receives them, they can be sent to the appropriate destination. Error detection is performed only at the destination.

Fast-packet technology using the frame relay approach can handle massive amounts of packets efficiently because it can allocate bandwidth dynamically. Northern Telecom has added frame relay to its DMS SuperNode, which enables carriers to offer frame relay as a service. U.S. Sprint was the first carrier to offer frame relay services the third quarter of 1991, with other major carriers following shortly afterward. Frame relay interfaces to bridges and routers are available from a number of vendors, including Newbridge Networks, Vitalink, and Digital Equipment Corporation. Frame relay offers a low-cost alternative to private circuits and is ideal for applications that don't need guaranteed up-time.

Frame relay offers greater efficiency than X.25 technology because it combines the statistical multiplexing and port-sharing functions found in X.25 packet switching with the high speed and low delay of time-division multiplexing.

As frame relay technology matures, new specifications from the frame relay forum as well as add-on services from carriers are making this technology more attractive for network managers who are developing enterprise networks that incorporate WANs. The frame relay forum has developed a switched virtual circuit (SVC) standard. Up until January 1994, frame relay communications services were limited to permanent virtual circuits (PVCs), which permitted communications only within certain geographic areas or business entities. With PVC, when ordering a circuit users must tell the carriers the point of origin and the virtual path's destination. The carriers must make any changes to the PVC. The SVC standard expanded frame relay's impact on wide area networks because it permits dynamic switching of frame relay connections across geographical regions in a seamless manner. SVC users have much greater control of communications.

Carriers are also actively improving the options available for frame relay. U.S. Sprint, for example, offers frame relay subscribers a zero committed information rate (0-CIR) option. This means you pay for frame relay's high speed without a guaranteed amount of bandwidth. As long as public frame relay is still a novelty, 0-CIR makes sense because of its price advantage. Because there's still quite a bit of frame relay capability available, users who opt for 0-CIR can still be assured of that service. A typical customer would access the network using a 64-Kbps line, and the odds are excellent that the frames delivered at the other end will still be close to the 64-Kbps rate, even if the 0-CIR option has been selected.

Switched Multimegabit Data Service

To confuse network managers even more, carriers offer switched multi-megabit data service (SMDS), which competes directly with frame relay. SMDS, unlike frame relay, offers connectionless data services that require no call setup or tear-down procedures. All attached nodes can be connected on the fly. SMDS's connectionless nature permits a virtual LAN to be created over a metropolitan area. SMDS service is available at several different speeds, ranging from 1.14 to 45 Mbps, but this service isn't available in all areas.

SMDS offers considerable flexibility in connecting many different sites where dedicated links would be prohibitively expensive. The service is ideal for smaller organizations with high-volume data needs who can't afford dedicated private networks. SMDS is designed for applications that consist of frequent but short transmissions, such as electronic mail or limited database access. Network managers can screen addresses to create their own virtual networks in which only authorized participants can receive data or send data. it's also possible to broadcast data to multiple recipients.

Cell Relay Technology

In the near future, It will also be possible to use fast packets with an approach known as *cell relay*. Cell relay is a very-high-speed switching system for public networks that is just starting to evolve. Unlike frame relay, cell relay relies on standard frames that don't change size regardless of the traffic load.

Cell relay has transmission speeds in the 150-Mbps range and is associated with two telecommunications standards, distributed queued dual bus (DQDB) and asynchronous transfer mode (ATM). You've already looked at DQDB earlier in this chapter, with the IEEE 802.6 specifications associated with a metropolitan area network. ATM transmits small packets over short internetwork hops. These packets, or *cells*, are routed by hardware platforms operating at very high speeds. You'll take a very close look at ATM technology in chapter 9.

Synchronous Optical Network (SONET)

In 1986 Bellcore proposed SONET, the synchronous optical network standard, to synchronize public communications networks and then tie them together via high-speed fiber-optic links. SONET defines a set of framing standards that determine how bytes are transmitted across links. What SONET offers is enormous bandwidth based on multiples of the base rate (OC-1) of 51.840 Mbps or one T-3 link. The U.S. originally proposed a 51.84-Mbps signal, while Europeans wanted a 34-Mbps signal. The two groups compromised at a basic rate of 155.52 Mbps or OC-3 to connect the two groups' signals. Table 7.1 provides the SONET hierarchy of transmission options.

Since SONET is built on the foundation of T-3, it's crucial that companies thinking of purchasing T-3 multiplexers in the immediate future receive a commitment from their vendors that the multiplexer is migratable to the emerging SONET standard. The device must have sufficient backplane bandwidth to support at least SONET's OC-1 and ideally enough bandwidth to support multiples of OC-3 for future growth. While the SONET equipment that's already available will be used mostly for public data networks, some companies, such as Apple, are already using SONET for their own private networks.

SONET's Phase II will provide operational and administrative support, including maintenance information on the data traveling over the links. Phase II will define the full seven-layer OSI protocol stack, including protocols associated with flow control to prevent data collisions. Among the carriers offering SONET are AT&T, MCI, and the regional Bell operating companies.

In the not too distant future, many industry experts forecast that telephone companies will offer customers dial-up 50-Mbps circuits using SONET rather than having to lease SONET lines. It's also likely that customers will be able to purchase fractional SONET service. Compared to the currently highest available dial-up rate of 64 Kbps, this additional bandwidth will make it economically advantageous to send video images on an as-needed basis without committing to a leased line.

Summary

Metropolitan area networks (MANs), defined by the IEEE 802.6 specifications, use a dual bus topology with data traffic traveling in opposite directions.

TABLE 7.1 The SONET Hierarchy

Optical carrier number	Transmission rate (BPS)	T-1s	T-3s
OC-1	51.84	28	1
OC-3	155.52	84	3
OC-9	466.56	252	9
OC-12	622.08	336	12
OC-18	933.12	504	18
OC-24	1244.16	672	24
OC-36	1866.24	1008	36
OC-48	2488.32	1344	48

The protocol used by MANs is distributed queue dual bus (DQDB) protocol. MANs are designed to handle city-wide communications.

Wide area networks (WANs) link together networks located in different geographic areas. Historically WANs have been limited by the relatively slow speeds available on dial-up analog phone lines. Recently, though, new technology in the form of T-1 lines provide a 1.544-Mbps bandwidth. T-3 lines provide a signal speed of 45 Mbps.

A relatively new development has been the availability of fractional T-1 service. Users can purchase only the fractional portion of a T-1 line they need. T-1 and fractional T-1 service can be used in conjunction with remote bridges and routers to form a seamless wide area network.

Packet-switched networks use the X.25 protocol. Public X.25 networks charge by the packet, so corporations with very heavy traffic sometimes install their own private X.25 networks. Hybrid X.25 networks represent the best of both worlds by using public networks where there's light traffic and private networks where they have heavy traffic.

Fast-packet technology is much more efficient than packet switching because of its ability to dynamically allocate bandwidth. Frame relay protocol is what makes fast-packet technology so efficient. The future might see the growth of cell relay technology, a fiber-optic approach that incorporates some of the same protocols associated with metropolitan area networks. Synchronous optical network (SONET) is a new standard for communications networks, with a transmission rate of 150 Mbps. In the near future, companies might be able to purchase fractional use of SONET services on an as-needed basis.

8

Integrated Services
Digital Network (ISDN)

In this chapter, you'll examine:

- ISDN's basic rate interface and primary rate interface
- The significance of signaling system 7 (SS7)
- The evolving 802.9 standard for integrated voice/data transmission
- Why broadband ISDN seems so attractive to network managers

For there to be an information superhighway, there must be high-bandwidth transmission at a reasonable price. Integrated services digital network (ISDN) is here; it's no longer a theory, but a high-speed digital network that can carry both voice and data information. In this chapter, you'll look at the different components or building blocks of ISDN, and also at an evolving IEEE 802.9 standard for integrated voice/data transmission that will be ISDN-compatible. Finally, you'll examine a broadband version of ISDN capable of transmission speeds over 100 Mbps. While some of this information might have been of interest at one time only to "telephone" people, today's network manager must know enough about these new technologies to be able to consider them within the context of designing enterprise networks.

What Is ISDN?

Integrated services digital network (ISDN) is an evolving set of international standards for connecting voice, data, and video equipment. With

ISDN, a user can carry on a voice phone call while simultaneously viewing video images or retrieving information from a computer. All these different forms of information can travel in a single ISDN interface circuit packet and be directed to an integrated voice/image/data terminal. The user can also select and change whether connections are to a private voice network or a public data network.

ISDN interfaces between local exchanges and end users could replace many of the links network managers currently use, including T-1 trunks, tie trunks, WATS lines, and traditional analog trunks. In 1976 the term *ISDN* appeared in the CCITT's "orange book" list of terms. It has taken so long to develop that skeptics have been known to refer to it as "I still don't need it" or "it still does nothing." Today, ISDN demonstrations at major telecommunications conferences are routine; some companies, such as McDonald's and Hardees, are already using ISDN.

ISDN still isn't present everywhere, and therein lies a serious problem for network managers trying to design a wide area network. The regional Bell operating companies (RBOCs) have different deployment schedules, so it's common to see one area of the country with much more ISDN penetration than another area. ISDN will be better than the alternatives only when it's available everywhere. At that time, presumably, the price of ISDN for linking branch offices throughout the country will be much less expensive than options such as leased lines. At this time, the first question you need to ask is whether ISDN is available at all sites to be linked into a wide area network. The second question is how expensive it is. There's no standard pricing, and each carrier has developed its own convoluted set of pricing algorithms.

The basic rate interface (BRI)

The basic rate interface (BRI) specifies a single access point into ISDN. Known as *2B+D*, BRI consists of two bearer channels and one data channel. Each bearer channel operates at 64 Kbps and is a clear channel, meaning there's no restriction on the format or type of information that passes through it. The data channel operates at 16 Kbps and is used for signaling and control information. The basic rate interface is also known as the *digital subscriber line (DSL)*.

The primary rate interface (PRI)

The primary rate interface (PRI) is used to connect multiple users to ISDN. Also known as the *extended digital subscriber line (EDSL)*, it will be used primarily to connect a PBX, LAN, or other multiuser switching device to an ISDN network. The North American standard, followed by the United States, Canada, Mexico, Japan, and South Korea, consists of 23 B channels of 64 Kbps each, and one D channel of 64 Kbps. The aggregate capacity is 1.544 Mbps, or the equivalent bandwidth of a T-1 facility. T-1 is intended to be the chief facility used with the North American standard for PRI. The European standard for PRI consists of 30 B channels and 1 D channel, for an aggregate capacity of 2.048 Mbps. Because of the greater capacity available under PRI, it supports an additional type of channel, H. Three types of H channels are specified:

- H0 (384 Kbps)
- 11 (1.536 Mbps)
- H12 (1.920 Mbps)

The North American standard incorporates H0 and H11 channels, while the European standard incorporates H0 and H12 channels.

ISDN Equipment

No one ever said that telecommunications jargon would be logical or easy to understand. The CCITT has arbitrarily defined several different types of ISDN equipment. *TE-1* equipment is ISDN-compatible and can be connected directly to the network. *TE-2* equipment isn't ISDN-compatible and requires an interface device known as a *terminal adapter (TA)*. A TA can convert signals from one international standard, such as RS-232C, to the ISDN standard. This provision for TE-2 equipment represents the CCITT's recognition that companies have so much money invested in existing non-ISDN telecommunications equipment that there's no way that ISDN will grow unless existing equipment can be integrated.

Network termination equipment can take two different forms under ISDN. *NT1* describes public switched network demarcation devices, such as a termination block or a registered jack. This equipment will have some built-in intelligence under ISDN because of the functions it must perform. *NT2* is the designation for customer-owned switching equipment, such as a PBX or a LAN. NT2 equipment can provide additional capabilities beyond NT1, such as call switching or concentration.

There are two additional types of ISDN equipment. *Line-termination equipment (LT)* is located within the local exchange company's or common carrier's network in situations where lines must be extended beyond the normal range of the central office. *Exchange termination equipment (ET)* terminates the digital subscriber line or extended digital subscriber line in the local exchange. ET can be characterized as central office equipment, important to the phone company but nothing you need to worry about.

Network Interfaces

Now that I've listed the various categories of ISDN equipment, the logical question is how these different types of equipment can be linked to an ISDN network. Figure 8.1 illustrates how interfaces link different types of equipment to an ISDN network. An R interface links non-ISDN-compatible equipment and terminal adapter equipment. The S interface links ISDN-compatible equipment and network terminal equipment. A T interface links customer premises equipment to an ISDN network, while the U interface ties together network termination equipment, exchange termination equipment, and line termination equipment.

ISDN in Action at the Hospital

Let's create a scenario to illustrate how the various ISDN components work together. Assume that a major hospital in Los Angeles uses several outside doctors as consultants. These consultants examine X-rays and then make their diagnoses. Normally a doctor is sent 12 X-rays, which are transmitted from the hospital to the doctors' offices using ISDN's basic rate interface. Figure 8.2 illustrates how the images are transmitted from the hospital's mainframe computer via NT1 equipment. This information travels to a Pacific Bell central office (CO) and over the ISDN network to the appropriate doctor's office, where it's received via an ISDN terminal. The doctor can study the X-rays and then send back an analysis that travels at ISDN's 64-Kbps rate.

Signaling System 7

ISDN requires a great deal of intelligence from the public switched network in order to format and transmit signals successfully. Signaling system 7 (SS7) consists of a series of recommendations from the CCITT that define the content and format of signaling messages under ISDN as well as the network design parameters necessary for transferring network control information.

There are three key components to SS7: service switch points, signaling transfer points, and service control points. Service switch points, usually

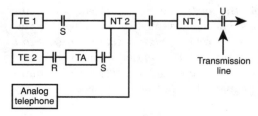

Figure 8.1 ISDN equipment types and interfaces.

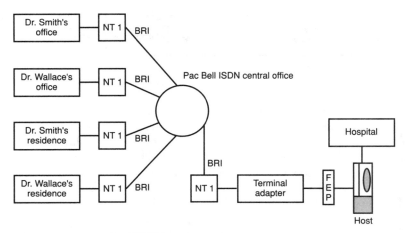

Figure 8.2 An example of ISDN in action.

owned by the LECs, receive call routing and handling instructions from a service control point. Signaling transfer points are packet switches that transfer information between an LEC network to a common carrier network. Finally, service control points are nodes containing computerized database records that provide control information to other nodes on the network. This control information could take the form of records of services to which each user subscribes or a list of toll-free 800 numbers. This information could be passed back to the appropriate node needing this information.

ISDN provides a number of services that require the intelligence available through SS7. Its use of out-of-band signaling conforms to ISDN's use of the D channel for signaling.

ISDN and the OSI Model

The D channel under ISDN uses the I.450/I.451 protocols at the network layer to recognize the type of packet as well as the type of message being transmitted. This layer covers the establishment, maintenance, and termination of calls through the network. Establishment includes setting up the call, selecting the type of service, and routing the call. Maintenance includes monitoring the call to ensure it isn't dropped or disrupted before normal termination. Termination refers to the orderly disconnection of a call.

As Figure 8.3 illustrates, the information field of an ISDN frame carries key network layer control information. The protocol discriminator field indicates which protocol is being used. The call reference field correlates the frame with the correct call. The message type field identifies a frame, which could take the form of a setup frame, information frame, acknowledgment

Protocol discriminator	Call reference	Message type	Information element ID	Length	Information	Information element ID

Figure 8.3 Frame structure of the ISDN information field.

frame, etc. The information portion of the frame also includes control data concerning its length and the sequence of the frames.

In ISDN's data link layer you'll find a number of protocols, including LAP-D, SAPI, and TEI. The LAP-D protocol, a subset of the OSI model's HDLC protocol, handles the flow of frames through the D channel and provides information for detecting and controlling data flow as well as recovering errors. LAP-D is a variation of the LAP-B protocol used on the B channel. LAP-B allows only one logical link across an interface, while LAP-D allows multiple logical links by permitting the address field to change from frame to frame. The flexibility to change addresses explains why a single D channel can control many B channels. Figure 8.4 reveals the format this protocol frame takes.

Another key data link layer protocol under ISDN is the service access point identifier (SAPI), which is used to multiplex packet, signaling, and management information over a single D channel, while the terminal endpoint identifier (TEI) is used to multiplex several logical D channels into one physical D channel. At ISDN's physical layer, the I.430 and I.431 protocols are designed to handle BRI and PRI respectively. Figure 8.5 indicates the protocols found under ISDN at each corresponding layer of the OSI model.

Integrated Voice & Data Under ISDN: IEEE 802.9

For several years the IEEE 802.9 committee has been working on a set of specifications for a standard to support voice and data transmitted over a single network. The committee wants to ensure that this new standard will provide IEEE 802 MAC services while being ISDN-compatible.

The 802.9 standard will cover communication between an integrated voice/data terminal (IVDT) and a LAN, where the IVDT's access unit (AU) provides all required services. It will also cover the situation in which the services required for an IVDT to serve as a gateway to a backbone network provide IVD services, perhaps as an 802.6 MAN using FDDI.

Flag	Address SAPI	Address TEI	Control	Control	Information	Frame check sequence	Flag

Figure 8.4 The ISDN LAP-D frame.

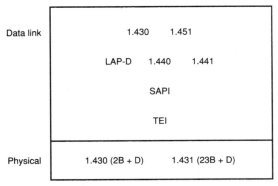

Figure 8.5 ISDN and the OSI model.

The 802.9 standard must be able to provide multiple access to a single wire. It does so by using time-division multiplexing and assigning a block of time to each channel. The IEEE 802.9 standard will include a P-channel protocol at the MAC level designed to handle packet-mode or burst data. The link access procedures on the D channel (LAPD) protocol used by ISDN at the MAC level is also used by 802.9 networks. This protocol ensures that 802.9 integrated voice/data networks will be able to use ISDN D channel control information.

Figure 8.6 illustrates the 802.9 MAC frame format. The flag consists of a 01111110 bit pattern that indicates a frame's start and end. The service identification (SID) field indicates what kind of service is provided on the P channel; it could, for example, designate ISDN's LAPD protocol or the 802.2 LLC protocol. It might even designate the X.25 packet-switching protocol. The field control (FC) field specifies whether the frame is used for control purposes or to transmit information. It also indicates the frame's priority.

The destination address (DA) and source address (SA) fields are self-explanatory. The protocol data unit (PDU) contains MAC control information or user data from a higher layer. Finally, the frame check sequence (FCS) uses the CRC-32 error-checking scheme.

The 802.9 specifications are now a standard. The success of 802.9 is dependent on the success of ISDN. If ISDN is successful, 802.9 will provide the "missing link" for a LAN's MAC layer protocols and the protocols associated with ISDN. It will be the glue that makes simultaneous voice and data transmission possible between a LAN and an ISDN integrated voice data terminal.

Flag	SID	FC	DA	SA	PDU	FCS	Flag

Figure 8.6 The 802.9 MAC frame format.

ISDN in Action at McDonald's

The common carriers have each selected key customers as beta sites for ISDN trials. McDonald's serves as an example of how these customers are planning to use ISDN. McDonald's Corporation has developed plans in conjunction with AT&T and Illinois Bell for a global ISDN network. The company has indicated that it will use a combination of 2,100 BRI and PRI lines at its corporate headquarters to link it with 40 domestic field offices, more than 8,000 U.S. stores, and 3,000 overseas stores. A major advantage of ISDN for McDonald's is that it will be able to consolidate 21 existing telecommunications networks into one. A user at any ISDN terminal will be able to access database information anywhere on the network.

McDonald's has indicated that one reason it's opting for ISDN is the services available on such a network. The company plans to use the ISDN electronic directory, including caller identification, message waiting, and message retrieval. Caller identification displays the number of the person calling from within the company and that person's name. Message waiting alerts users when a message has been left for them, and message retrieval enables users to examine a series of waiting messages by scrolling through them.

Broadband ISDN

Many industry experts have begun to say openly what they've been thinking for quite a while—that narrowband ISDN might have arrived too late. ISDN's individual 64-Kbps channels and its aggregate 1.5-Mbps might be insufficient for the bandwidth required by LANs and enterprise networks. Broadband ISDN (BISDN) raises the bandwidth threshold to over 600 Mbps, which might be necessary for large-scale transfer of high-resolution video along with voice and data information. The CCITT has identified three H, or high-speed, channels to be used in conjunction with BISDN. Table 8.1 lists the CCITT ISDN hierarchy of H channels.

TABLE 8.1 The CCITT ISDN Hierarchy of H Channels

H channel	Gross bit rate*
H11	1.544
H12	2.048
H21	34.368
H22	44.736
H4	139.264

*The gross bit rate includes control data overhead.

Do You Need ISDN?

ISDN is just one of several different communications options a network manager must consider. Companies that today depend heavily on Centrex, large private networks, and public data networks and plan to establish global links with international branches will embrace ISDN because it will save them a substantial amount of money.

Companies that already have committed heavily to LANs and have bypassed technologies such as VSATs and microwave might not need ISDN. Narrowband ISDN's 64-Kbps channels and 1.544-Mbps bandwidth aggregate might be inadequate for companies that have WANs that must carry substantial amounts of information. Broadband ISDN is still too far in the future to serve as a practical solution to these WAN bandwidth requirements.

Summary

ISDN's basic rate interface (BRI) consists of two 64-Kbps bearer channels and one 16-Kbps data channel. The primary rate interface (PRI) is also known as 23B+D because it consists of 23 B channels of 64 Kbps each and one 64-Kbps D channel. ISDN equipment contains several different kinds of interfaces. TE-1 equipment is ISDN-compatible, while TE-2 equipment is not ISDN-compatible and requires a terminal adapter. NT-1 equipment consists of public switched demarcation devices, while NT-2 equipment describes customer-owned switching equipment such as a PBX.

The IEEE 802.9 committee has developed a draft for a set of specifications for integrated voice/data transmission. This draft's provisions are compatible with ISDN and metropolitan area networks.

Broadband ISDN might be the answer for network managers who find narrowband ISDN inadequate for their wide area network needs. Broadband ISDN will have bandwidth in the 600-Mbps range, large enough to handle integrated voice/data/video transmission.

Chapter

9

Asynchronous Transfer Mode (ATM)

In this chapter, you'll examine:

- What ATM is and why it has attracted so much interest
- ATM's cell structure
- How ATM works as part of a local area network
- How ATM works as part of a wide area network
- The general growth pattern for ATM over the next few years

A network manager would have to be a hermit to avoid hearing the term *ATM*. In this chapter you'll examine asynchronous transfer mode (ATM) and see how it works and what unique features it can bring to a network. You'll also look at some major problems that still exist in linking the new world of ATM to the legacy LANs that currently exist. While the first ATM products have been developed for local area networks and workgroups, later products will be developed for PCs wide area networks. You'll see the reasons why the evolution of ATM will probably follow this pattern.

What Is Asynchronous Transfer Mode (ATM)?

Asynchronous transfer mode (ATM) is a switching technology that uses small, fixed-size cells. In 1988, the CCITT designated ATM as the transport mechanism it planned to use for future broadband services. ATM is asynchronous because cells are transmitted through a network without having to occupy specific time slots in a frame alignment, as is found in T1 frames. These cells are small (53 bytes), compared to variable-length LAN packets. ATM is a connection-oriented technology, in contrast to most LAN-based protocols, which are connectionless. A *connection-oriented* approach means that a connection needs to be established between two end points with a signaling protocol before any data transfer can take place. Once the connection is established, ATM cells are self-routing because each cell contains fields identifying its connection to the cell to which it belongs.

Transmissions of different types, including video, voice, and data, can be intermixed within an ATM transmission that can achieve speeds ranging from 155 Mbps to 2.5 Gbps. This transmission speed can be directed to a desktop user, workgroup, or entire network. Because ATM doesn't reserve any specific positions within a cell for specific types of data, its bandwidth can be optimized by allocating bandwidth on demand. The small, fixed size of ATM cells results in predictable throughput with short delays. Switching the fixed-size cells means incorporating the algorithms in silicon chips and eliminating delays caused by software. Another advantage of ATM is that it's truly scalable. Several switches can be cascaded to form larger networks.

Asynchronous transfer mode is based on the concept of two end-systems (terminals) communicating via a set of intermediate switches. There's the user-to-network interface (UNI) and the network-to-network or network-to-node interface (NNI). The UNI links an end-user device and a public or private ATM switch, and the NNI describes a link between two switches. Figure 9.1 describes an enterprise network using both UNI and NNI.

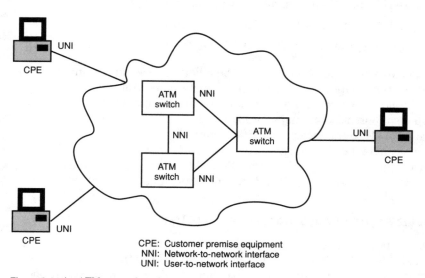

CPE: Customer premise equipment
NNI: Network-to-network interface
UNI: User-to-network interface

Figure 9.1 An ATM network and its interfaces.

In the case of public ATM service, when it comes, the UNI describes customer premise equipment's interface to a public ATM switch, which could be a stand-alone device, such as a workstation, or a private ATM switch. Specifications have been developed for public and private UNI. There are two public UNI interfaces, one at 45 Mbps and another at 155 Mbps. The DS3 interface is defined in an ANSI committee T1 draft standard, while the 155-Mbps interface is defined by the CCITT and ANSI standards groups. Three interfaces have been developed for Private UNI—one at 100 Mbps and two at 155 Mbps. It's likely that the international standard SDH/SONET 155-Mbps interface will become the interface of choice because it permits interoperability at both the public and private UNI.

Because ATM is a connection-oriented network, a connection between two end-points begins when one transmits a signaling request across the UNI to the network. A device responsible for signaling then passes the signal across the network to its destination end-system. If this system indicates it agrees to the connection, a virtual circuit is set up across the ATM network between the two systems. Both UNIs contain mappings so the cells can be routed correctly. Each ATM cell contains two fields, a virtual path identifier (VPI) and a virtual circuit identifier (VCI), that indicate these mappings.

Because each ATM cell contains the VPI/VCI value, even if an end system has two more virtual networks across a UNI, ATM cells can be interleaved between the two circuits and the data routed to the appropriate system upon its arrival.

The ATM Forum

At the heart of the push for international standards for ATM is the ATM Forum. Formed at the Fall 1991 Interop show to develop standards at the LAN level for this new technology, the group has grown from four companies to over 350 vendors of public, private, and computer networking equipment. The ATM Forum is divided into two main groups, the Technical Committee and the Marketing, Education, and Awareness Committee. The Technical Committee breaks into a number of subcommittees that are developing ATM specifications in the following areas:

Signaling

This committee develops specifications for the components of user-to-network interfaces. Such interfaces must be able to provide multipoint-to-multipoint connections, multicasting, and adding and dropping connections to an existing call. Point-to-multipoint service is a desirable feature for many developers because it would make it possible to implement multiparty videoconferencing, shared workgroups, and distributed routing updates. The hope is that ATM multicasting will permit multiparty videoconferencing without a bridge. This committee is also developing interfaces for frame relay internetworking.

B-ICI

The B-ICI committee is working on a broadband intercarrier interface that will encompass traffic management and performance parameters. It's also developing specifications for usage monitoring for accounting purposes and a definition of switched virtual circuit cell relay and frame relay services.

UTP-3

This committee is developing specifications for low-speed ATM transmission over category 3 unshielded twisted-pair wire. ATM can run at speeds ranging from 51 Mbps to 1.2 Gbps. While ATM is associated in most people's minds with high-speed transmission, IBM has developed and is selling ATM chips with 25-Mbps speeds. Among the companies licensing this chip are TranSwitch Corporation. Integrated Device Technology has also developed its own 25-Mbps ATM chips.

Why would a network manager want a 25-Mbps ATM? IBM argues that this speed represents a logical upgrade for token-ring users now restricted to 16 Mbps. The relatively low cost associated with this upgrade would make it economically attractive, particularly because it can be used with type 3 cabling already installed at most locations.

There are some clear negatives to this low-speed ATM. Assuming that this topology would be associated with workgroups rather than higher-speed backbone transmissions, why choose a 25-Mbps ATM rather than a 100-Mbps Ethernet or token-ring version sold by IBM and Hewlett-Packard? There's no clear reason why a network manager should opt for this topology. It's yet another example of different parts of IBM offering competing solutions for the same workgroup problem. In this case, low-speed ATM loses because it's still far more expensive than 100-Mbps token-ring hubs.

Traffic

This committee is developing specifications on traffic management. The document they produce will also cover traffic descriptions and contracts as well as burst-level control techniques.

Private NMI

This committee is developing specifications for a network-to-network interface. It must deal with such issues as physical layer discovery, routing protocols, traffic management, flow control, signaling procedures, administration, and security.

Network management

This committee is developing specifications for network management. Its major concern is that whatever schemes it develops must be interoperable with existing network management programs.

Service aspects and applications

This committee is working on the very crucial issue of LAN emulation and interconnection. It must ensure that companies with legacy LANs, such as Ethernet and token ring, are able to interoperate with ATM switches via high-speed, low-cost terminal emulation. The committee is also looking at videoconferencing over ATM and both frame-relay-to-ATM and SMDS-to-ATM internetworking.

Testing

This committee is developing specifications for testing ATM systems via diagnostics. It's charged with developing the means for ATM diagnostic test access, interoperability and conformance tests, and the establishment of an interoperability laboratory.

The ATM Protocol Reference Model

Just as the OSI layered protocol model describes communication between two computers over a network, the ATM protocol model describes how two end-systems communicate via ATM switches. As shown in Figure 9.2, the key layers that need explanation are the physical, ATM, and ATM adaption layers. That part of the layered architecture used for end-to-end or user-to-user data transfer is known as the *user plane (U plane)*. The control plane defines higher-level protocols used to support ATM signaling, and the management plane (M plane) provides control of an ATM node and consists of two parts: plane management and layer management. The plane management function manages all other planes and the layer management function is responsible for managing each of the ATM layers.

The physical layer

The physical layer defines the physical interfaces and framing protocols associated with ATM. This layer is segmented into two sublayers—the transmission convergence (TC) sublayer and the physical medium dependent (PMD) sublayer. The reason for sublayers in this ATM architecture is to de-

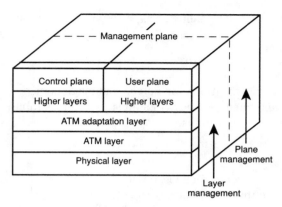

Figure 9.2 The ATM protocol reference model.

couple the transmission from the physical medium to permit a variety of physical media.

The TC concerns itself with adaptation to the transmission system, defined as the reception of cells from the ATM layer and their packaging into appropriate format for transmission over the PMD. The TC also handles cell delineation, cell scrambling/descrambling, and cell rate decoupling. Cell delineation is the extraction of cells from the bit stream received from the PMD. The function of cell rate decoupling is to insert/suppress idle cells to or from the payload in order to provide a continuous flow of cells. Finally, the TC generates and verifies the header error check (HEC). It calculates the HEC from the bits received and checks it against the HEC value of the received cell. If there's a match on consecutive cells, then the TC assumes correct cell boundaries. If there's no match for many successive cells, the TC knows that the correct cell delineation isn't yet found.

The ATM layer and ATM cells

The ATM layer performs four basic functions. It multiplexes and demultiplexes cells of different connections. These connections are identified by their virtual circuit identifier (VCI) and virtual path identifier (VPI) values. It also translates VIC and/or VPI values at the switches or cross connects, if required. The ATM layer is also responsible for extracting/inserting the header before or after the cell is delivered to or from the adaptation layer. Finally, this layer handles the implementation of a flow control mechanism at the universal network interface (UNI), using the general flow control (GFC) bits of the header.

Figure 9.3 describes the ATM cell format, which consists of a five-byte header and a 48-byte information field. The cell size is deliberately small so

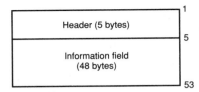

Figure 9.3 The ATM cell format.

there's very little delay or latency in the delivery of information, particularly synchronous traffic.

The ATM cell header

Figure 9.4 describes the ATM cell header fields found in cells going across the UNI and the NNI. The four-bit generic flow control (GFC) field is used only across the UNI to control traffic flow and prevent overload conditions. This field isn't defined across the NNI, and the corresponding bits are used for an expanded virtual path identifier (VPI) field.

The virtual path identifier (VPI) field is used to identify virtual paths. Consisting of eight bits across the UNI and twelve bits across the NNI, the field isn't yet defined by either the CCITT or ATM Forums. The virtual circuit identifier (VCI) field is 16 bits long. End devices assign a value to the VPI and VCI fields when requesting a connection to an end system.

The payload type identification (PTI) field consists of three bits and is used to identify the payload type carried in the cell, as well as to identify control procedures. Future specifications developed by the ATM Forum will

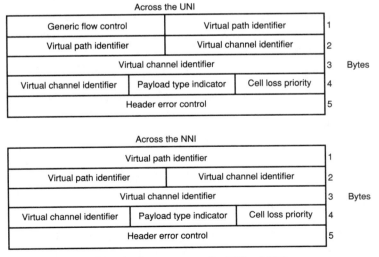

Figure 9.4 ATM cell header formats across the UNI and NNI.

designate the setting of one bit to indicate congestion, a second bit for network management, and the third bit to indicate an error condition.

The cell loss priority (CLP) field is a single bit that indicates a cell's loss of priority. This bit is set to 1 when a cell can be discarded due to congestion; if a switch experiences congestion, it will drop cells with this bit set. This results in giving priority to certain types of cells carrying certain types of traffic, such as video, in congested networks.

The header error check (HEC) is an eight-bit cyclic redundancy code that's calculated over all fields in the ATM header. This type of error checking can determine all single-bit errors and a number of multiple-bit errors. Error checking is very important in ATM operations because an error in the VPI/VCI could cause corruption of the data flow of other virtual circuits.

General ATM Operation

ATM requires that a connection be made between two end points before information can be exchanged. An end point on a network sends a signal across the UNI to the network requesting a connection to another point. The network sends this request to its destination point, where it's interpreted. If this node accepts the request for a connection, a virtual circuit is established across the network. A VPI/VCI is set on both UNI and on the intermediate nodes on the network. Two end or switching points can be linked via a virtual channel link. The VPI and VCI fields of the ATM cell header contain the routing information required.

Figure 9.5 illustrates how this process works. End point A requests connection to point B. The ATM network assigns value P to the VCI of A, and value Q to point B. Node A will use P in outgoing information, and B will use Q for incoming information. Lookup tables are set up throughout the network. The same process is followed for the reverse direction. The first switching node that receives a cell from A will consult its lookup table to find out where the cell should be switched to and what value the outgoing VCI should be assigned to. This process is repeated until it arrives at B.

How does the ATM layer function when the ATM node is an end system? The AAL layer provides it with information. When the ATM layer exchanges a cell stream with the physical layer, it inserts this information as well as the required parameters in the header fields, including the crucial VPI/VCI values. If it has no information to transmit, it fills the information field with idle cells. The ATM layer is also responsible for controlling the quality of service for each circuit, a value that's negotiated when circuits are established. Among the parameters that are negotiated are the peak and average data rates, the acceptable delay, and the loss rate.

The ATM layer's operation is even less complicated for a switch. The ATM layer under this circumstance receives an ATM cell on one port and uses the VPI/VCI value to determine to which port to forward the cell. It then for-

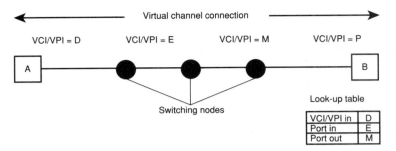

Figure 9.5 An ATM virtual channel connection.

wards the cell to the appropriate port, changes the VPI/VCI to reflect the cell's routing, and transmits the cell to the physical layer of that port.

The Four Classes of ATM Traffic

A major advantage of ATM in an enterprise network environment is its ability to handle a variety of different types of traffic. Current specifications define four different classes of traffic that can be handled. ATM is designed to handle the following different classes of traffic:

Class A. Constant bit rate (CBR), connection-oriented, synchronous traffic (e.g., uncompressed voice or video)

Class B. Variable bit traffic (VBR), connection-oriented, synchronous traffic (e.g., compressed voice and video)

Class C. Variable bit rate, connection-oriented, asynchronous traffic (X.25, frame-relay-like services, etc.)

Class D. Connectionless packet data (LAN traffic, SMDS, etc.)

Let's take a closer look at how these levels of service are handled by the ATM adaption layer (AAL) protocols. The AAL provides service to the higher layers that correspond to the four classes of traffic. Figure 9.6 describes the nature of the AAL.

It's important to distinguish between the service offered to the higher layers by the AAL and the type of traffic carried by the AAL protocol across the network. Type 1 protocol is designed to handle the needs of constant bit traffic (CBR) network services. An example of this type of traffic would be a T1 (1.544-Mbps) signal used to transfer bits at a constant rate between two points. While AAL type 1 typically offers class A service and carries class A traffic, the type of service and the type of traffic carried by the AAL

	Class A	Class B	Class C	Class D
Timing and relation between source and destination	Required		Not required	
Bit rate	Constant		Variable	
Connection mode	Connection-oriented		Connectionless	
AAL types	1	2	3/4,5	3/4

Figure 9.6 AAL layers.

don't always have to correspond. Class A is associated with connection-oriented, constant-bit traffic. A higher layer provides the AAL with a fixed number of bits at a regular frame rate. The AAL delivers the bits to their destination at the same bit rate. Generally, the ATM layer uses priorities to ensure that the ATM cells are sent and received at the same rate across an ATM network.

Class B traffic is connection-oriented, but permits a variable bit rate. Data is passed to the AAL from the higher layers at fixed intervals, but the amount of data might vary from transmission to transmission. If the amount of data exceeds the capacity of a single cell, then the data is segmented into multiple cells and reassembled at the data's destination.

Class C traffic is connection-oriented, variable-rate data with no timing relationship between source and destination. Packet-switched networks such as X.25 and frame relay require this type of service. This is a very key part of the AAL because it will probably be the heart of wide area network transmission. The problem is that it's a very complex layered protocol to define. Much of it is based on the IEEE 802.6 protocol for metropolitan area networks. There's now a complete AAL standard for class C, type 5, the simple subset of AAL 3/4.

The AAL type 3/4 protocol has been developed to support SMDS connectionless service. It's compatible with the IEEE 802.6 protocols that support SMDS. This is a bit tricky because SMDS uses variable-sized frames that must be delivered error-free. The AAL type 3/4 protocol can handle this problem by using a transmitter to segment each SMDS frame into fixed-size segments and then place them into ATM cells. A receiver is also needed to reassemble the SMDS segments into the original frame. The entire process is known as *segmentation and reassembly*.

ATM Protocols in Action

Fore Systems is one of the leading ATM switch manufacturers. Figure 9.7 shows how the company's ATM software permits network management in the form of an SNMP agent. It also facilitates transmission of legacy LAN data via proprietary LAN emulation.

ATM LAN Emulation

For ATM to play a major role in enterprise networks, there must be a set of specifications for ATM LAN emulation. This software permits low-cost bridges to inexpensively and efficiently bridge data in legacy LAN packets, such as Ethernet and token ring, to high-speed native ATM ports by mapping a legacy LAN MAC address to an ATM address. For this emulation to be successful, an ATM media access layer must be defined to be totally indifferent to upper-layer protocols while still being compatible with Ethernet, token-ring, and FDDI media access layers. The going has been slow in this area at the ATM Forum because many ATM switch vendors have developed their own proprietary legacy LAN emulation software and consider it to give them a competitive advantage; as a result, many of these vendors have been slow to disclose some of their software secrets to their competing fellow members of the ATM Forum.

Development of terminal emulation within the ATM Forum is proceeding in two phases. The first phase will develop specifications so that anyone's end system will be able to talk to anyone else's LAN emulation service. The second phase will include distributed address resolution. The long-term terminal-emulation goals of the ATM Forum are to make existing LAN applications ATM-aware and to enable new applications to take advantage of ATM's quality of service features. To accomplish this particular goal will re-

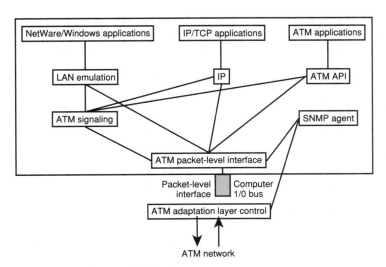

Figure 9.7 Fore Systems' ATM software.

quire the development of native ATM application program interfaces (APIs) so that applications can have raw access to ATM.

ATM Network Management

There are two primary functions associated with network management of an ATM network: internal connection management and external network management. Internal connection management is responsible for call setup, call routing, address resolution, and management of switched virtual circuits and permanent virtual circuits. The internal network management function must be contained within the ATM switch and ATM adapters because call setup must be performed continually and as quickly as possible by the ATM network.

Connection management software is also responsible for performing automatic network configuration, which permits discovery of end nodes, switches, and other ATM-attached devices. This approach eliminates the trouble and expense associated with moves and changes in traditional LANs. The connection management software also monitors performance of every port and every virtual path, trunk, and channel and collects statistics on usage and errors.

Much work still needs to be done. At this time, for example, there's no ATM equivalent of status signaling for use by the internet protocol's address resolution protocol (ARP). This means that there's no way an IP user can check a PVC's status or a network can inform a user of PVC status change. The real significance of this limitation is that, when conditions change, a router can't determine what is connected at the other end of the PVC.

While terminal emulation is one of the major issues associated with ATM on local area networks, an equally serious issue is the fact that LAN protocols aren't designed to run efficiently in an ATM switch environment. Novell has plans to overhaul its protocols to work more efficiently with ATM on LANs. In addition to working on ATM LAN emulation, Novell has announced plans to develop a more efficient version of its internetwork packet exchange (IPX), which will offer native ATM transport services. IPX was designed originally for broadcast, connectionless networks; this means that it doesn't support the virtual connections used in ATM's connection-oriented technology.

A second problem with the current version of NetWare is that its servers use Novell's service advertising protocol (SAP). This protocol periodically broadcasts the availability of its services across a local area network. In a connectionless environment, clients simply pull the message off the network, but that's not the case in a connection-oriented network where these periodic broadcasts would tie up the limited number of virtual circuits supported by workgroup switches.

External Network Management of ATM Networks

While several ATM vendors have developed their own proprietary network management software, the real future will be in SNMP-compliant software. The first stage now underway is for ATM vendors to provide SNMP MIBs for the various elements within their products. These MIBs permit network managers to use standard SNMP management systems to monitor and configure ATM networking equipment. The MIBs are likely to show up first as vendor-specific MIBs. A second stage in ATM network management development will be standard MIBs with vendor-specific extensions. Vendors at that point will then provide ATM network management software that will run on existing management platforms but offer even more advanced capabilities.

The Internet Engineering Task Force (IETF) is now working to define an ATM management information base in conjunction with the ATM Forum. At the same time, the International Telecommunication Union's Telecommunications Standardization Sector and the American National Standards Institute (ANSI) are working to develop network management specifications for public networks. When the work of these two very different groups eventually converges, SNMP will most likely manage private ATM networks, while the common management information protocol (CMIP) will be used within and between public networks.

ATM Private Network-to-Network Interface

Network managers today are reluctant to mix and match ATM switches. A private network-to-network interface (P-NNI) specification will make it more likely that ATM switches from different vendors will interoperate.

In order for P-NNI to permit this level of interoperability, it must deal with the tricky issue of routing. An ATM network can contain thousands of connections, some at constant bit rates, others at variable bit rates, and still others at available bit rates. The routing decisions for all these connections must ensure quality of service to a wide variety of traffic types.

The quality of service (QOS) includes such variables as cell loss ratio, cell delay, and delay variance. An ATM network must negotiate these parameters depending on the different types of traffic and the traffic requirements for applications such as voice, data, and video. The network must establish traffic prioritization and congestion management. Congestion management is a major problem for ATM because of the need to enforce network parameters, such as committed information rate (CIR), cell loss ratios, delay and delay variance, and burst size and duration—all of which must be managed in a nonblocking environment.

To make matters even more complex, switched virtual circuits (SVC) must be able to dynamically negotiate each of these parameters with the network in real time. Congestion management is responsible for seeing that each switched virtual circuit performs up to the conditions established in the QOS parameters, previously described. There are several proposals on how congestion should be handled if it occurs. One proposal, known as forward explicit congestion notification (FECN), has the ATM switch that's experiencing congestion set a bit in the ATM headers of all cells passing through it. Destination stations recognize the congestion bits and send a special cell back to their sources to reduce the rate of transmission.

How ATM Will Grow

The first ATM products available were used in workgroups, particularly composed of high-performance workstations used for high-bandwidth applications such as computer-aided design and engineering. The lack of standards for LAN emulation caused these first products to perform only as stand-alone switches rather than as part of large networks. Also, these first switches were much too expensive to be much in demand on the PC desktop. The major use for these switches over the next few years will be as corporate backbones. Intelligent hubs are likely to be linked to the ATM backbone switches to handle local traffic. The hubs are likely to have the routing function integrated into their backplanes.

ATM as a backbone

One of the major functions ATM will perform in enterprise networks will be that of a high-speed, highly scalable backbone. One early example of how this process works is SynOptics' EtherCell stand-alone Ethernet-to-ATM switch. The switch includes 12 10Base-T Ethernet ports and an ATM port that can be multimode fiber or twisted-pair wire. This particular switch performs ATM signaling for Ethernet hosts and is linked to an ATM switch that provides the backbone function. Figure 9.8 reveals how the Ethernet-to-ATM switch works in conjunction with an ATM switch.

ATM can be integrated directly into an intelligent hub to serve as a network backbone. Hughes LAN Systems' enterprise hubs incorporate an ATM cell-switching fabric as an integral part of their architecture. The ATM backplane coexists with legacy LAN segment Ethernets, token rings, and FDDI. The ATM backplane supports all types of traffic (AAL1 through AAL5), and permits configurations that include point-to-point, point-to-multipoint, and multipoint-to-multipoint connections. Hughes handles the LAN emulation requirements by offering an ATM router access module. This is a full-slot module that provides both an ATM backplane and an optional external UNI port via an ATM media module to Wellfleet router modules. This router module routes legacy LAN traffic to ATM through ATM adaption layer 5 (AAL 5).

Irvine-based LANNET has come up with an innovative way to integrate ATM functionality without having to upgrade existing equipment. The com-

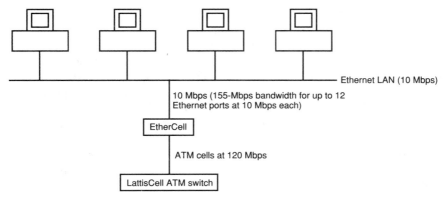

Figure 9.8 The EtherCell linked to an ATM switch.

pany's partnering with ATM switch vendor General DataComm will result in a module that permits LANNET intelligent hubs to function as bridge/routers for local LAN segments tying into an ATM backbone. This bridge/router will provide an interface between the intelligent hub's 1.2-Gbps protocol-independent, cell-switching bus and a General DataComm ATM switch. There are two intriguing aspects to the approach that LANNET has taken. Besides being a cost-effective way to link an intelligent hub to ATM without having to move to a different chassis, this strategy also means that the products will be found in the corporate wiring closet and not in the basement as part of some backbone switch device connecting servers.

ATM on the desktop

ATM is unlikely to arrive on the PC desktop in any large numbers before 1995 because it will take that long for the price of these adapter cards to become reasonable enough for mass purchase. Plans are already underway to ensure that ATM will be able to become an integral part of local area networks. Because Novell's NetWare dominates the LAN environment, it's no surprise that several companies are working actively with Novell to ensure that their ATM adapter cards will include NetWare drivers. Two key LAN areas where ATM will play an active role on the desktop are as a part of a remote-access server and as a switch used in conjunction with Novell's multiprotocol router software. The first products for ATM at the desktop are likely to be servers. A very significant portion of network traffic consists of the multitudes of messages sent from server to server on a network.

Switched Virtual Circuits

ATM permits both permanent virtual circuits (PVCs) and switched virtual circuits (SVCs). The advantage of SVCs are that this feature enables a switch to dynamically establish connections. PVCs must be set up in advance. The ATM Forum has adopted the Q.2931 specification for setting up SVCs, as part of the ATM Forum's UNI 3.0 set of specifications. From a customer's perspective, it's much better to purchase a switch that supports Q.2931 rather than a proprietary protocol. PVCs are used when only a few connections are needed. Because PVCs are defined in terms of their end points, a network could vary with each exchange. Devices linked by a PVC must maintain tables that track all connections. For this reason, workstations joined by a PVC would require tables listing every other workstation on the network; this is clearly not practical and helps explain the purpose of an SVC.

The basic signaling capabilities supported by the UNI 3.0 agreement include the following:

- Switched channel connections

- Point-to-point and point-to-multipoint switched channel connections

- Connections with symmetrical or asymmetrical bandwidth agreements

- VPCI/VIP/VCI assignment

- A statistically defined out-of-band channel for all signaling messages

- Error recovery

- Public UNI and private UNI addressing formats for unique identification of ATM end points

- A client registration mechanism for exchange of addressing information across a UNI

Figure 9.9 shows how public and private UNI switches can be linked using this set of specifications.

ATM on a Wide Area Network

The CCITT published its I.432 standard defining the synchronous optical network's (SONET's) 155-Mbps and 622-Mbps rates as the physical layer for broadband ISDN back in 1990. SONET will form the worldwide telecommunications standard for fiber-optic transmission systems linking together ATM switches. In 1992 the American National Standards Institute (ANSI) provided draft standard for a physical layer BISDN user-to-network interface based on the DS3 (45-Mbps) rate and format. Because DS3 rates are widely tariffed, it's relatively easy for the carriers to develop BISDN services for ATM now and assume that the applications needed to drive SONET's higher speeds will take some time. When ATM wide area network services become available, it will be possible for ATM LANs to support direct connections into an ATM wide area network through DS3 and SONET switch interfaces.

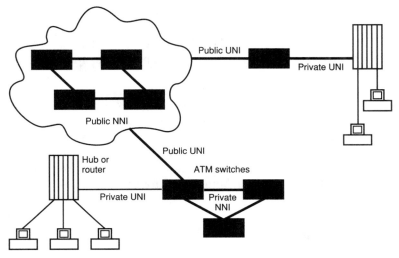

Public UNI

Private UNI

Public NNI

Public UNI

Hub or
router

ATM switches

Private UNI

Private
NNI

Figure 9.9 An ATM wide area network.

To make this interoperability possible, the ATM Forum has been developing a user-to-network interface (UNI) based on CCITT standards. The UNI specifications will support not only interoperability over SONET but also interoperability over multimode fiber at 100 and 155 Mbps, as well as over DS3 connections at 45 Mbps. It's likely that full ATM interoperability between LAN and WAN won't be available until around 1997.

There are significant additional problems involving wide area network ATM not found in local area network ATM that are likely to help account for the slow implementation of ATM by the carriers. Perhaps the major problem is that many of the vendors developing switches for wide area networks still view this emerging market the same way they view the traditional PBX market; they assume that proprietary features are much more likely to sell a product than interoperability. Some of these vendors have clearly been dragging their feet on the issue of interoperability—in sharp contrast to the LAN vendors, who have been the model of cooperation at the ATM Forum.

A potentially even more serious problem for wide area ATM is that ATM LANs don't have to concern themselves about the differences among committed information rates, burst rates, or variable bit rates. These specifications differ from carrier to carrier. Today carriers are supporting only class A (constant bit rate) and class C (variable bit rate) service. They don't yet support class B (variable bit rate, constant timing, associated with the transmission of packet video) and class D (connectionless service associated with SMDS).

Pricing as well as the class A features offered vary widely among carriers as they begin to offer rudimentary ATM wide area service. AT&T's InterSpan

ATM service offers a flat rate, while MFS Datanet offers a combination of flat rate and usage-sensitive pricing. Virtually all carriers now offer their own proprietary flow-control protocols. Class A features offered vary, from T1 granularity with infinite change to 64-Kbps granularity with 24-hour notice.

What might hinder the development of these classes of WAN service is the lack of applications for wide area networks that require such high-speed transmissions. Many MIS directors are beginning to look seriously at the possibility of using ATM's T3 speeds because they might prove much more cost-effective. Today there are no multimedia, video-intensive networking applications that require greater than T3 speed. The carriers are looking for customers willing to pay premium prices for high-speed wide area ATM service, but there might not be enough before 1997 for these services to really grow.

Key Features for an ATM Switch

The following sections list a number of ATM switch features that a network manager might want to include in an enterprise network. Because ATM technology is so new and products are as yet unproven, the best advice for the present is to find a vendor whose switches have the best combination of features and price and then buy all switches from a single vendor. For the next few years there likely won't be true ATM interoperability because each ATM vendor now has its own proprietary connection management software, naming services, flow control techniques, and congestion management software.

Switched virtual circuits (SVCs)

As discussed earlier in this chapter, a network that includes a number of PCs and high-performance workstations would create real problems for an ATM switch that was limited to permanent virtual circuits (PVCs). The routing tables to be maintained by the various switches on this network would prove unwieldy and could create delays in transmission. SVCs would eliminate the need to maintain these complex routing tables and improve network efficiency.

Network management

Some ATM vendors now include proprietary network management software, while others provide SNMP compatibility so any standard SNMP-compatible network management program can integrate ATM management into overall enterprise network management. Fore Systems' ForeThought, for example, uses SNMP protocol to work in conjunction with the company's ForeRunner switches, which include an integral SNMP agent. This combination of hardware and software means that the switches can be managed by any SNMP network management software, such as OpenView or Sun Connect.

The ATM Forum has developed an interim local management interface specification (ILMI), which supports SNMP MIB II. The SNMP MIB information supports inventory management, including switch hardware serial number, switch hardware version, software version, and network module count. It also includes configuration information that ranges from data on VPs originating on the switch, VCs passing through the switch, the switch address, port number, and IP address of the port connected device. Other network management functions supported by SNMP MIB II include bandwidth utilization (ATM cells transmitted/received, maximum bandwidth, current port bandwidth utilization, etc.) and errors, including those in physical layer framing and VCIS out of range. Because the ATM Forum still has a long way to go in developing network management standards for ATM switches, the ability to transmit management information to an SNMP-compatible network management program is very desirable.

Some vendors, such as NEC, are promising to provide support for the ATM Forum's specification for private NNI (P-NNI). This protocol will permit the interconnection of ATM switches within private networks, and appropriate network management software will be crucial for effective use of this protocol.

DS-3 and SONET WAN interfaces

ATM switches can't function in a vacuum. Many companies who have made the strategic decision to purchase ATM switches have done so for two primary reasons. The first is to provide a high-speed network backbone, and the second is to move toward a long-term strategy of integrating LANs and WANs. ATM switches with DS-3 and SONET interfaces will be able to transmit data from LAN to WAN once the wide area set of specifications for ATM have been finalized.

Downloadable software

Because ATM is in such a state of flux, the ATM Forum is likely to revise most of the current ATM specifications. An ATM switch that's capable of receiving downloaded software from any network workstation will be easily upgraded to reflect new standards.

Switch transit delays

Some ATM switches will be purchased specifically for transmitting real-time video. For this type of transmission, it's essential that the switch transit delay be less than 10 microseconds to maintain acceptable video performance.

Aggregate nonblocking switching capability

Particularly for backbone performance, it's essential that network managers size the switches they purchase to ensure that there's enough nonblocking switching capability to handle network data traffic. Blocking switches typically offer higher bandwidth than their nonblocking counterparts and try to minimize the amount of blocking. SynOptics' Lattiscell, for example, uses a technique called *randomized traffic distribution* to ensure that calls are rarely blocked.

Many switches are now available with up to 10-Gbps aggregate nonblocking switching capability. This overall switching capacity will continue to meet the growing need for greater capacity. Take, for example, a 16-port ATM switch with a 2.5-Gbps fabric. Because of ATM's use of a contentionless time-division switching fabric, each port always gets a time slot $\frac{1}{16}$ of the 2.5-Gbps fabric, or a full 155 Mbps. Because the traffic from each port arrives at a maximum rate of 155 Mbps, the traffic is never blocked from entering the switching fabric, hence the term *nonblocking*.

Diagnostic software

Vendor-supplied software to enable network managers to monitor ATM local area equipment and wide-area connections from a customer site will be very desirable for network management. This software should also be able to monitor ATM service usage at the public ATM UNI interface for utilization and error statistics. This data will enable network managers to troubleshoot ATM equipment, resolve potential discrepancies with billing, and determine network availability as well as error rate information.

Flexible port speeds

The ability of an ATM switch to support multiple port speeds will enable network managers to assign the appropriate port speed for each network user. If a user's needs change, then the port speed can be changed. There has been a good deal of interest in T-1 speeds for some ATM ports, for example.

Router interfaces

Today's routers connect to ATM networks via hardware or software. The ATM Forum has developed a hardware specification known as the data exchange interface (DXI), which describes the way LAN packets are to be recognized by devices like DSUs. This approach enables an ATM device to know where to find the addressing information within a LAN packet and how to divide this packet into cells. A router can also link to an ATM switch via frame relay if the switch is equipped with a frame relay interface.

Most routers cannot currently perform ATM cell segmentation and re-assembly by themselves. A few router manufacturers, such as Retix, offer these hardware interfaces.

ATM and Today's Routers

ATM poses a long-term threat to router manufacturers, but the first and second generation of ATM switches are likely to be used primarily as backbones, and companies will continue to use existing routers on the periphery of networks. These routers will use ATM as their transport service. There are currently no direct links between ATM and router networks. Routers will continue to handle multiprotocol routing and internetworking of various network segments for the short term, but routing functionality will probably be situated within the ATM switch and large-scale routers will eventually disappear.

One harbinger of things to come is the ATM connect edge router from Agile Networks. This family of routers brings ATM switching together with routing and packet-to-cell conversion capabilities. This type of product has been defined by some as an edge path adapter (EPA) because it links connectionless LANs to connection-oriented ATM networks. Some EPA products will reside at the outer edge of a campus network and provide access to ATM WANs and public services, while others will be mounted in wiring closets and eliminate the need for campus routers.

NetEdge's edge router sits at the edge of the LAN and ATM network, distributing intelligence to access points and providing end-to-end services and SNMP-based network management. One advantage of such a product is that the distributed intelligence of edge routers ensures greater reliability, security, congestion control, and billing and performance optimization. Because it's designed specifically to work with ATM switches, its OC-3 UNI ATM interface supports 30MB of throughput, compared to conventional routers that provide only 7MB to 12MB of throughput.

ATM as a Part of Virtual Networks

Enterprise networks by their very nature will include all kinds of topologies, including ATM, and will permit the creation of virtual networks. This means users anywhere on the network can be switched from segment to segment while maintaining a virtual workgroup address. Such an approach would simplify the network manager's job of handling individual moves and changes. The limitation of traditional routers is that they require the physical attachment of a network segment to each router port. NetEdge System's ATM Connect is the first of what should be a whole new class of router products. This product uses a network, protocol-based virtual networking architecture rather than a physical, connection-based architecture.

Stations on an ATM network can be interconnected at the full bandwidth of the media without needing a traditional LAN partitioned via bridges and routers to handle performance problems. Each connection carries only the traffic required on that specific link.

ATM Strategy: IBM and AT&T

IBM and AT&T offer the most comprehensive ATM strategy today. An examination of how each company's various LAN and WAN ATM components fit together will help tie together the various ATM topics you've seen in this chapter and also provide a preview of what to expect from ATM in the near future. IBM refers to its comprehensive plan as broadband network services (BBNS), which includes access services, transport services, and control point services.

BBNS access services support the emerging protocols and services offered by a BBNS network. IBM has developed access agents that provide access services by translating the external supported protocols (HDLC, IP, etc.). External nodes connecting to BBNS can communicate with its access services via native protocols without knowledge of BBNS features. These access services are smart enough to understand and interpret the external services or protocols. They can also locate the target resources by performing address resolution. They maintain and take down connections across a network in response to connection requests, as well as manage bandwidth to ensure fairness among users.

BBNS transport services provide transport across the network for user traffic generated at the edges. They perform transmission scheduling and hardware-based switching with multicast capability. BBNS transport services support ATM switching as well as automatic network routing, and match appropriate service to QOS classes.

BBNS control point services control, allocate, and manage network resources, including bandwidth reservation, topology updates, and group management support for multipoint connections. This set of services identifies new resources and users automatically, supports multiple virtual private networks and multiple virtual LANs on the network, and assists in establishing and maintaining multipoint connections.

IBM's comprehensive ATM strategy includes both 25-Mbps and 100-Mbps adapters, ATM workgroup concentrators, ATM switches, and switch network control management that runs on a NetView/6000. IBM's ATM plan includes expanding its 8260 intelligent hub to include ATM switching capability, developing integrated bridging and LAN emulation servers, and developing an ATM 100-Mbps adapter for its 3172 mainframe controller, as well as an interface for its intelligent hub.

At the heart of all these products is IBM's "switch-on-a-chip" design, developed at the IBM Research Laboratory in Zurich. This chip features 16 input ports and 16 output ports, all of which operate simultaneously. This single chip can drive more than eight gigabytes of aggregate throughput and work in conjunction with multiple chips to produce higher throughput.

Figure 9.10 shows IBM's vision of ATM in a campus environment. It encompasses ATM at the desktop, LAN links, and a mainframe link and includes ATM switches, adapter cards, network management software, and the appropriate interfaces to other systems. A 100-Mbps LAN adapter supports NetWare as well as OS/2. LAN emulation is provided by an IBM proprietary approach using a LAN-to-ATM bridge. Protocols supported include IEEE 802.3 (Ethernet), IEEE 802.5 (token ring), IEEE 802.2 (logical link control), NETBIOS, IPX, and IP. The company promises to support ATM Forum LAN-emulation standards when they're firmed up via a software upgrade.

IBM has committed to support the ATM Forum's user-to-network interface (UNI), which permits ATM devices such as servers and workstations to connect to an ATM switch. It will also support whatever network-to-network interface (NNI) that emerges to permit ATM switches to connect to each other.

While IBM is a key member of the ATM Forum, its ATM product line includes at least one key nonstandard offering. The company's Turboways 25 adapter provides 25-Mbps ATM service to the desktop. IBM argues that there are several reasons why there's a need for low-speed ATM. IBM argues that the 25-Mbps version of ATM is cost-effective because it gives companies

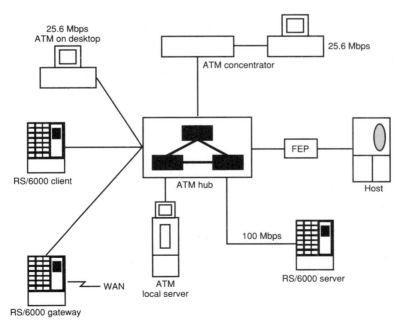

Figure 9.10 IBM's vision of an enterprise ATM.

investment protection by enabling them to use existing technology. The ISA adapters transmit over voice-grade unshielded twisted-pair wire, the cabling already installed in most establishments. The current technology that produces 25-Mbps ATM permits a very low cost per port at the desktop. IBM offers concentrators to share the per-port cost over multiple devices.

Because the ATM Forum hasn't gone along with IBM on the issue of a 25-Mbps ATM standard, the company has chosen to go its own way on this issue. It has established a consortium of 16 network equipment and semiconductor suppliers to support the technology. As mentioned earlier in this chapter, Hewlett Packard is a key supporter of IBM strategic direction with this product.

AT&T's ATM strategy is built around its proprietary ATM Phoenix chip. Using this chip as a building block, the company has created a family of products, including the UniverCell workgroup, departmental and backbone switches, and an intelligent hub. The Phoenix chip provides ATM cell switching and includes the following features: adaptive routing for fault tolerance, four levels of priority, a feedback mechanism for congestion control, and multicasting capability.

The SmartHUB XE ATM switch is an intelligent hub that contains a pluggable ATM switching module. The module can be managed via simple network management protocol (SNMP). Figure 9.11 illustrates how an AT&T Phoenix ATM switch can be linked to the SmartHUB's ATM module as well as to LANs, ATM workstations, and file servers.

AT&T believes that there will be a demand for departmental ATM switches that can be linked together via FDDI translation. These switches will offer dedicated 10-Mbps Ethernet and 16-Mbps token-ring interfaces, as well as low ATM, T1, and T3 speeds. Figure 9.12 shows the role of a departmental ATM switch in an AT&T environment.

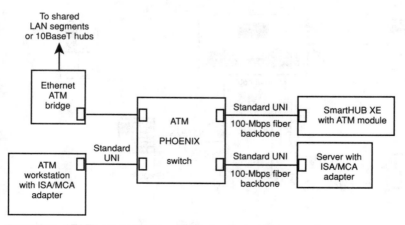

Figure 9.11 AT&T's comprehensive ATM network strategy.

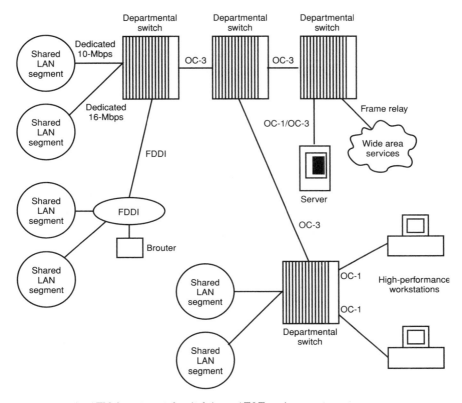

Figure 9.12 An ATM departmental switch in an AT&T environment.

Summary

Asynchronous Transfer Mode (ATM) is a high-bandwidth transport technology that transmits 53-byte fixed-size cells at speeds ranging from 155 Mbps to 2.2 Gbps. ATM switches are now being used to form high-performance workstations in workgroups. The major market for ATM switches will be as corporate network backbones. One of the major problems facing the ATM Forum is to develop a set of specifications for LAN emulation, a way to seamlessly link legacy LANs with ATM switches. At this time there are only proprietary solutions.

10

The Branch Office

In this chapter, you'll examine:

- The differences between remote control and remote node links
- The role of the remote access server
- The advantages of routing software to create wide area networks
- How hubs and routers can be integrated to create WAN links
- The importance of point-to-point protocol and the IPXWAN protocol

The linkage of branch offices to corporate headquarters via LAN communications will grow more than 33 percent annually between 1994 and 1997, according to market research leader Computer Intelligence InfoCorp. There are a bewildering number of options for network manager planning to link branch offices with corporate headquarters to form enterprise-wide networks. There are also different types of remote users with very different needs. A large branch office that needs constant linkage with the home office requires a far different product than a nomadic salesperson who occasionally needs to access a branch sales office or a LAN at corporate headquarters. In this chapter, you'll examine some of the different types of remote access products and consider the environments in which they're most appropriate. Before looking at the specific needs of branch offices, it's useful to review the basics of remote communications.

A Remote Communications Primer

Because remote communications offers so many different options and because the terminology is so obtuse, let's spend a few moments reviewing some of the key types of remote communications.

Data codes

Before meaningful communications can take place, data must be coded into a form that can be transmitted and then decoded by the recipient. In 1963 the American National Standards Institute (ANSI) developed the American Standard Code for Information Interchange (ASCII), which has become a standard for PCs and even for some larger computers. While there are some variations, generally seven bits are used to describe a character, number, or symbol, with the eighth bit reserved for parity error checking. (You'll return to the subject of error checking a bit later in this chapter.)

IBM's mainframe computers use the data coding scheme extended binary coded decimal interchange code (EBCDIC), in which all eight bits of a byte are used to describe a character, number, or symbol. Clearly, the two schemes are not compatible, which means that some sort of translation must take place for remote communications to occur between stand-alone microcomputers or microcomputers on LANs and mainframe computers.

Asynchronous communications

The world of PCs and PC LANs uses asynchronous communications, also known as *start/stop communications*. The eight bits of transmitted information is followed by a signal representing a stop bit. This stop bit lets the receiving unit know that the transmission is complete. Timing between the receiving and sending computers is resynchronized, a start bit signals the beginning of another byte, and the next byte of information is transmitted. This is a very inefficient method of transmission because of the overhead associated with stop bits.

Synchronous transmission

Larger computers tend to use synchronous transmission, which consists of two computers synchronizing their clocks to determine when transmission will begin. Then a large block of data is transmitted, followed by bits that provide control information, including error checking. The receiving computer needs this information to unpack the data and interpret its various fields. Unlike asynchronous transmission, where there's only 80% efficiency (two bits out of every ten are used for stop bits), synchronous transmission is over 95% efficient.

Modem Standards for Remote Communications

Remote communications requires a knowledge of modems and some modem standards because such a high percentage of remote communications still consist of modems and dial-up lines rather than the more expensive digital leased lines. Many of the country's largest businesses still use 9,600-bps modems despite the availability of faster modems (14.4, 19.2, and even 28.8 Kbps).

One reason so many 9,600-bps modems are still being used is the availability of data compression. By compressing data that's transmitted, more information can flow with fewer data bottlenecks. Unfortunately, modem manufacturers tend to offer proprietary data compression schemes, so it's very risky to purchase modems from different manufacturers. Microcom, for example, uses a technique called adaptive Huffman encoding, which assigns a token to each eight-bit pattern. The modem keeps a frequency table and then assigns the shortest token to the most frequently occurring character. A maximum of 2:1 compression can be achieved in some cases.

The CCITT has issued its V.42bis recommendation for data compression, which uses a technique known as BTLZ. A 4:1 compression can be achieved with this approach. A string of two to four characters is replaced with a codeword. A dictionary of more than 512 text strings can be built to contain these codes. The data is encoded and then transmitted using the codewords. The recipient modem looks up the codes in the dictionary and decodes them into characters. The V.42bis specification requires that, if two systems use different sized dictionaries, they agree to communicate with the smaller dictionary for data compression.

Microcom Networking Protocol (MNP)

Because of politics, U.S. modem manufacturers lost out in trying to get the CCIT to adopt the microcom networking protocol (MNP) rather than the V.42bis standard. Despite this setback, MNP plays a very prominent role in U.S. remote communications. MNP is a de facto standard for both error correction and data compression, and offers several different classes of support. Over 15 million modems currently use MNP. Let's take a quick survey of the different classes of MNP available.

MNP class 1

The lowest class of support, MNP 1 uses an asynchronous byte-oriented half-duplex method of exchanging data. This means that the transmitting and receiving computers alternate using the transmission line, much the same way as a citizen's band (CB) radio. After data has been transmitted, the recipient must transmit either an acknowledgment that the data has been received correctly or a negative response that the data needs to be re-sent. Most of today's modems no longer function at this low level.

MNP class 2

This level provides asynchronous byte-oriented full-duplex service, which means that data can flow in both directions at the same time. This class of service is also virtually obsolete today.

MNP class 3

This class adds greater efficiency by providing synchronous bit-oriented full-duplex transmission. The overhead of start and stop bits is eliminated and the user sends data in asynchronous format to the modem, which it turn communicates with a second modem via synchronous transmission.

MNP class 4

This class adds adaptive packet assembly, which alters the size of packets in which data is packaged and transmitted between modems based on the quality of the physical link. If the line quality is good, then larger packets are transmitted and greater efficiency is obtained. A second technique, data phase optimization, removes repetitive control information from the data stream to create more efficient packets. This class provides approximately 120% protocol efficiency compared to class 3 transmissions.

MNP class 5

This class adds data compression to achieve a net throughput efficiency of approximately 200 percent. The compression algorithm analyzes the data to be transmitted and adjusts the compression parameters accordingly.

MNP class 6

This class adds two additional techniques. Universal link negotiation en-ables a single modem to operate at a full range of speeds between 300 and 9,600 bps, depending on the maximum speed of the modem on the other end of the transmission. Two modems begin at a common slower speed and then negotiate the use of a higher speed. A second technique, statistical du-plexing, permits modems to simulate full-duplex service on a half-duplex modem connection.

MNP class 7

This class improves on the data compression offered in class 5. An average 300% improvement over a non-MNP modem is possible. An adapted Huffman encoding scheme is used. There's also a second type of data compression that supports V.42bis data compression, which can provide up to a 400% improvement in efficiency.

MNP class 9

Yes, there is no class 8. Class 9 improves the efficiency with which modems acknowledge that messages were received and retransmits information following an error.

MNP class 10

This class sits on top of the CCITT set of specifications, such as V.42bis for data compression and the various standards for modem modulation (V.22bis, V.32, and V.32bis). Therefore, it supports all the international standards and also offers the proprietary MNP algorithms.

There are five value-added MNP features in this class. Negotiated speed upshift begins the modem handshake at the lowest possible speed and then upshifts to the highest possible speed. Robust auto-reliable mode means that modems using this feature establish a reliable link during noisy class set-ups by making multiple attempts to overcome channel interference. Other classes of MNP support make only one attempt. Dynamic speed shift is a feature that continually adjusts speeds throughput the connection in response to prevailing line conditions. The combination of downshifting and upshifting of modulation speeds is designed to ensure reliable support throughout the life of the connection. Aggressive adaptive packet assembly is a feature that improves link and transmission performance under adverse conditions. Packet sizes vary from 8 to 256 bytes. Rather than lower classes of MNP protocol that begin with the largest size packets and then decreases the size if bad line conditions exist, MNP 10 protocol starts with the smallest data packets and then increases packet sizes as conditions permit. Finally, dynamic transmit level adjustment (DTLA) maximizes performance on cellular connections. It starts at 1,200 bps and determines a modem's ideal transmission level for prevailing line conditions. It then monitors line conditions once the cellular link is made and uses transmission level statistics to adjust the transmission levels to optimize data throughput.

Modem Pooling

Because very few users need to use their modems 100 percent of the time, it's much more efficient and cost-effective in an enterprise network environment to share modems in a modem pool, much the same way managers often share the services of secretaries in a secretarial or word processing pool. Modem pooling requires intelligent hardware and software to provide adequate monitoring and control. The software should be able to place modem requests into a queue and allocate them to modems as they become available. In addition, the modem pooling software must be able to handle different communications programs on different PCs.

Error Checking

Transmission over a LAN is relatively error-free compared to remote communications because line noise isn't really an issue. For remote communications it's crucial that the transmitting and receiving workstations be able to check for errors and determine when data packets need to be retransmitted. The asynchronous communications so common in PC LAN communications generally codes data in an ASCII format. ASCII uses seven bits of information to describe a control character, a letter, or number. The eighth bit is designated as a *parity bit*.

This type of error checking is often referred to as *vertical redundancy checking (VRC)* because one bit at the end of each byte of information is checked. Parity checking can use an even or odd parity. When even parity is used, all eight-bit units (bytes) have an even number of 1-bits. This means that, if a character or number's ASCII code consists of an odd number of 1-bits, an extra 1-bit is added at the parity bit position to ensure that an even number of 1-bits are transmitted. Similarly, if transmitting and receiving workstations have agreed on an odd parity error checking scheme and the five-bit representation of a character or number has an even number of 1-bits, then an extra 1-bit is added to the parity position to ensure that an odd number of 1-bits is transmitted.

No error checking scheme is perfect. The danger to using a parity checking scheme is that it can't protect against a situation in which a number of bits are damaged, with the net result that the odd or even status of the byte is unchanged. In this case, no error is identified and the wrong number of character is transmitted and received.

A second error checking approach, horizontal redundancy checking, checks for errors after an entire block of information has been sent. Finally, a longitudinal redundancy check (LRC) adds a block check character (BCC) at the end of a block of data. The first bit of each message viewed horizontally has a parity bit associated with it. There's a also a parity bit associated with each subsequent bit of the byte. Thus, each row of bits in all characters has a bit at the end that provides either odd or even parity. These bits form the block check character.

There's a more sophisticated form of error checking, cyclic redundancy checking (CRC), which is usually associated with mainframe and minicomputer transmissions of data. This scheme uses a complex algorithm consisting of dividing an integer by a prime number (a number that can't be divided evenly) and noting the remainder. This remainder is calculated at the receiving end of a data transmission and compared with the remainder calculated at the transmitting end. If the two numbers don't match, an error in transmission is noted and a request is made to retransmit the data.

XModem, YModem, and kermit

XModem is a very simple error checking scheme that's commonly used in the PC LAN world of remote communications. The receiver sends a NAK signal (negative acknowledgment character) and then the transmitter begins sending information in 128- byte blocks followed by a checksum. This checksum consists of a total of all the 128 bytes in a message. If the receiving PC determines that the checksum displays the correct number and that the sequential block number is one more than the previous block it has received, it then sends an ACK (acknowledgment) signal back to the transmitting PC. The transmitting PC then sends the next block of information. Should an error be determined, the receiving PC sends a NAK signal, which results in the transmitting PC resending the block of information.

There are some limitations to XModem. While it's adequate for most PC communications, it isn't sophisticated enough for communications among large computers, which transmit very large blocks of data unsuitable for its approach. XModem protocol is also limited to computers able to send and receive an eight-bit byte. Some communications channels are limited to seven bits per character because the eighth bit is designated for a special control function. These channels cannot use XModem. An equally serious problem for the network manager is that the XModem error checking process cannot be automated. It requires manual operation at both ends.

YModem is another error checking protocol that can handle batch transfers of information. Because multiple files are transmitted, information about each file is sent before the batch is transmitted. AN ASCII STX character at the beginning of the block is followed by 1,024 bytes (rather than 128 bytes under XModem). A version of YModem known as ZModem adds some additional features, such as the ability to restart the transfer of a file from the point it was interrupted rather than requiring that the entire file be retransmitted.

Kermit is an error checking protocol named after a character on *Sesame Street*. It transmits blocks of data but doesn't require an eight-bit byte block or packet. The default kermit packet is 80 characters long and ends with an end-of-line character, such as a carriage return. This protocol is particularly valuable because it can transfer information among a wide range of computers rather than being limited to computers that use ASCII format. It's available for over 100 different computers, from microcomputers to supercomputers.

Linking to Corporate Headquarters

There has been a revolution in the way branch offices are connected to corporate headquarters. In the late 1980s companies began replacing IBM front-end processors, controllers, and 37xx communications processors with bridge-routers. There was considerable cost savings because these routers, which cost between $20,000 and $30,000, were replacing equipment that often cost more than $100,000, but the price of the routers still was too prohibitive to permit replacement of all mainframe equipment. Many branch offices still used dumb terminals to dial into corporate headquarters.

Many of these same companies began replacing mainframe equipment at their corporate headquarters with local area networks. The need to link branch offices to these LANs became more acute because of the e-mail running on the corporate LANs. The concept of an enterprise network is built on the premise that all parts of the network are linked together closely to enhance communications. E-mail is as crucial as collaborative computing and the sharing of key files.

By early 1993, 3Com began offering low-cost routers designed specifically for branch offices. Major router vendors, including Cisco and Wellfleet, soon followed. At the same time, major intelligent hub vendors, such as Synoptics and Cabletron, began offering remote communications within their hubs. The branch office has emerged as a major area of LAN growth in the 1990s because there are still so many of these offices not yet linked to corporate headquarters.

What Do Branch Offices Require?

What types of features are important to network managers who purchase branch office solutions? Cost is obviously a consideration, particularly when dozens of identical configurations are required. A bank seeking to link 50 branch offices obviously wants the lowest-priced solution that will do the job because of the multiple units that must be purchased. Flexibility in the specific routed protocols is also important. Customers don't want to pay for protocol support they don't need, but they want to be assured that additional protocols can be added without having to upgrade. "Plug and play" has become a cliche in the computer industry, but it's an important consideration for most customers with multiple sites. Ease of installation is crucial because most branch offices don't have highly skilled network engineers.

Ease of remote management is also crucial. The vast majority of corporations haven't yet implemented comprehensive network management, but these same companies want to be able to manage their remote sites immediately.

Types of Remote Access

There are three different methods of connecting a remote user or branch office with remote users to a corporate LAN. *Terminal emulation* is a technique that uses software to link a user at remote terminal across a wide area network to another computer as if it were a locally attached node. *Remote control* is an approach that permits a remote user to seize control of a local PC on a corporate LAN. *Remote node* uses a remote access server serving as a kind of traffic police officer to enable a single remote user or a remote LAN filled with users to communicate with a distant corporate LAN. Let's take a closer look at these three methods of remote communications.

Terminal emulation

Terminal emulation is a very limited way for remote users to access a computer system. Historically, it has worked reasonably well with mainframe and minicomputer applications, but it isn't a valid option when the remote user wants to access a LAN because the remote user is limited to the functions provided by a terminal device. Because the focus on this book is on LAN communications over a wide area network, I won't discuss this type of remote communication any further.

Remote control access

Remote control access enables a remote user to dial into a LAN and seize control of a PC on that LAN. Major software packages that offer this type of remote service include Microcom's Carbon Copy, Norton-Lambert's Close-Up, and Synamtec's PCAnywhere. The processing speed and power of a LAN's PC that's being controlled determines the speed of the remote control session. This PC's network shell processes all LAN packets. It communicates over the LAN with its server at normal LAN speeds. Keystrokes typed on the remote PC are sent to the PC on the LAN that's controlled. This host PC in turn transmits any changes in its screen to be displayed on the remote PC's screen. The files and applications being used remain on the LAN rather than being downloaded to the remote PC. Only keystrokes and screen images are transmitted from host to remote PC, over a modem at speeds that generally range between 2,400 and 57,600 bps. Windows-based applications, because of their intensive screen updates, generally require data compression in order to avoid overly sluggish performance.

Remote control as a method for accessing LANs is declining in popularity for a number of reasons. One reason for this loss of popularity is that essentially two PCs are tied up to perform one job—the local PC and the remote PC that's controlling its functions. A second problem with this approach is that the remote control program requires users to learn a dif-

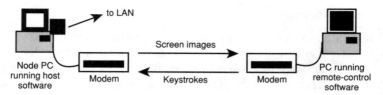

Figure 10.1 A remote control session.

ferent interface than is currently found on the LAN. Perhaps the major reason for the declining market share of this type of program is that it really doesn't handle Windows-based applications very well. The more programs that offer graphical user interfaces, the more difficult it will be for this type of remote access to offer acceptable transmission speeds. Figure 10.1 illustrates how a remote control session functions.

Remote nodes

Remote node software lets a remote PC function as a full-fledged user of a LAN. The user interface is identical to that displayed by a local PC on a LAN. In addition to software for the remote PC client, a remote node server is also required. Known as a *remote access server*, this server can be a modem with built-in software or a PC server running remote node software.

Usually a network manager makes sure that remote PCs have all the executable files (*.EXE) required for a session so the files don't have to be transmitted over the wide area network link. This approach is particularly important when accessing Windows programs because of the amount of data that otherwise would have to be transmitted to support the graphics-oriented user interface.

As a full-fledged LAN user, the remote PC has a network address. In the case of a NetWare LAN, a server would convert the network packets from their normal IPX protocol format into a format compatible with RS232 serial communications for transmission over a wide area network.

While a local LAN server can serve as a repository for all data files as well as application programs, the remote PC must have a license for the software package being accessed. Otherwise, the remote PC can't be loaded with the *.EXE files; the files must be accessed from the LAN server. Such an arrangement isn't often satisfactory because of the additional transmission time required to load all required files. Figure 10.2 illustrates a remote node session.

Remote Control or Remote Node?

While remote node is gaining popularity at the expense of remote control, there are times when remote control is perfectly adequate and, in fact, the preferred approach. If database programs with massive files must be accessed, remote control would be preferable because the processing of these large files would take place on the local LAN's host PC and only the required files would be sent over the LAN. For very expensive programs, a remote control approach means you wouldn't have to buy additional software licenses. This monetary savings must be balanced, however, with the requirement that a local PC must be tied up and accessed remotely.

The growth of client/server software will accelerate the trend toward remote node software because such software will process large data files on local LAN servers and transmit only required records to the remote PC.

Different Types of Remote Node Service

To further complicate matters, there are several different types of remote node service that use remote access servers. These remote access servers can take the form of router software running on PCs with appropriate hardware, bundled hardware (including internal modems) and software, stand-alone routers, routers integrated within hubs, and hub cards integrated within routers. Confusing, isn't it? Let's examine the options and see how they differ.

Dial-up LAN-to-LAN router software

One type of remote node service is the use of dial-up LAN-to-LAN routers, which take the form of sophisticated software installed on a PC that serve as a type of remote access server. Users are unaware that there's any difference between local and remote communications because the communication is transparent. All users have network addresses whether they're local or remote users.

Rockwell International's NetHopper is an example of this type of product. NetHopper supports IP over the PPP protocol, a couple of standards that network managers are starting to require to ensure some degree of interoperability. There's a limited amount of bandwidth available for communications, so NetHopper keeps LAN diagnostic messages, such as NetWare's "keep-alive" packets, from being transmitted over a dial-up line.

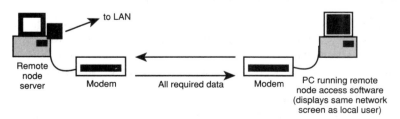

Figure 10.2 A remote node session.

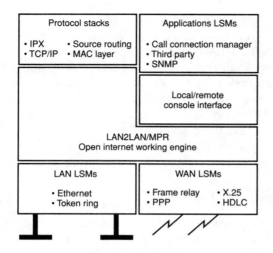

Figure 10.3 Protocols and functions in Newport Systems' LAN2LAN software.

Other dial-up LAN-to-LAN routers include Telebit's NetBlazer and Newport Systems' LAN2LAN. These products differ in the options they offer, including the type of network supported (Ethernet, token ring, etc.), the number of modem ports supported, and the ability to upgrade to faster links, such as T-1 lines.

Figure 10.3 shows the architecture of Newport Systems' LAN2LAN multiprotocol router software, including the protocols supported. Notice that the product supports a number of wide area network protocols as well as Ethernet and token ring's LAN protocols. This figure shows the software installed in a NetWare file server, which provides local and remote routing when configured as an external router. When installed on a NetWare file server or runtime, such as NetWare for SAA or NetWare Connect Services (NACS), this software permits users to interconnect NetWare LANs without having to purchase additional PC hardware. Figure 10.4 shows branch office users with access to all systems connected to the wide area network, including an AS/400, an Ethernet LAN, and an Ethernet LAN.

Dialing the Branch Office: The Nomadic Road Warrior

Many branch offices are sales offices. While salespeople need to be able to dial into corporate headquarters from the branch office, their major use of remote communications will likely be when they communicate with their offices from the road or home. Remote access equipment to handle the nomadic road warrior includes a communications server, a modem or modem pool, and both server and client software to make communications possible. One example of such software is Novell's NetWare Connect Services. Connect is a NetWare loadable module (NLM) that enables remote DOS-based PCs as

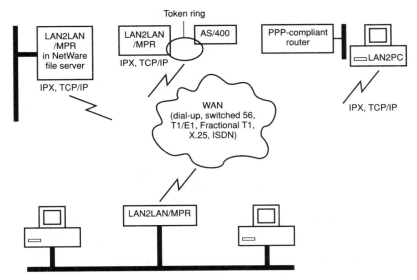

Figure 10.4 Newport Systems' LAN2LAN dial-up software in action.

well as Macintosh clients to dial into NetWare networks. Other examples include DCA's Remote LAN Node (RLN), TechSmith RemoteNB (RNB), and Stampede Technologies' Remote Office (RO). Figure 10.5 illustrates NetWare Connect Services in action.

There are major differences among these remote access communications products. Network managers evaluating these products might look for such key features (when appropriate for their own network environments) as the ability to support Windows as well as a Windows/NetWare environment, integration of server security and authentication within an enterprise environ-

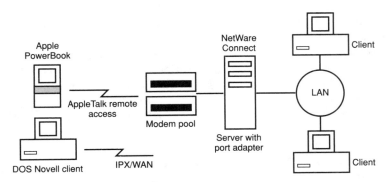

Figure 10.5 Nomadic users dialing in via NetWare Connect.

ment, ability to kick off users on a per-port basis, and detailed statistics and activity logging. Network managers must also be aware that the throughput of these software products depends on their underlying method of operation. A product such as Stampede Technologies' Remote Office functions as a bridge, while Novell's NetWare Connect is a router. A bridge product will provide greater throughput, but it won't be able to find alternative routing paths should a primary path go down.

There are some practical considerations as well. These products often differ widely in their support for asynchronous communications boards. NetWare Connect supports products from most board companies, including DigiBoard, Computone, Hayes, Gateway, Newport Systems, and Star Gate Technologies, while other software products support only a single board vendor. Ease of installation is also important, particularly when dozens of users are involved. Some programs require each remote user to load the required drivers manually or to write a batch file to automate the remote login process. Other programs provide a menu-driven interface. Other features that help differentiate remote access communications server software are:

- NOS versions supported
- Client memory requirements
- Host memory requirements
- Maximum port speed
- Maximum number of incoming lines
- Restriction of number of invalid consecutive login attempts
- Remote management via remote control
- Protocols supported

The impact of AT&T NetWare Connect Services

AT&T and Novell have jointly announced that AT&T's NetWare Connect Services will use Novell's NetWare Directory Service to provide intra- and intercompany users with secured access to applications as well as other networked resources. The service will support both Novell's IPX and the Internet's TCP/IP protocols. The first application to be supported is Network Notes; AT&T's servers will run Lotus Development Corporation's Notes groupware software.

The major advantage users could enjoy from NetWare Connect Services would be the seamless linking of distant NetWare LANs over a wide area network. Users wouldn't need to know where a server was located; the directory services would make the network connection transparent to the user.

There has been some controversy about this service because it uses Novell's proprietary technology rather than open industry standards for directories. The service will interoperate with X.400- and X.500-based services via Novell's directing mapping utility. (Chapter 11 discusses the advantages of X.400 and X.500 services.) As Figure 10.6 illustrates, this service will permit user access via wireless links, AT&T dial-up services, and AT&T frame relay service. The product will offer such features as security, user registration and profiles, and network management.

Bundled remote access server

Modem companies have seen their margins on modems tumble as the products became commodities. Many of them have begun to sell remote access servers. These products usually bundle the hardware and software required for remote access. Shiva's LANRover/Plus, for example, offers eight V.32bis modem modules that supports dial-in by DOS and Windows clients using Novell's IPX protocol as well as TCP/IP and NetBEUI protocols. Shiva includes software that allows you to manage multiple LANRover units as if they were a single box. This product also supports mixing and matching analog and digital lines. An ISDN module is offered as an option. The major advantage of Shiva's approach is that combining modems in the remote access server box saves money and space and permits easier management.

Micom's LANexpress server is an example of a remote access server with even more features. This product supports simultaneous remote control and remote node access on a single connection, and supports the Novell IPX, TCP/IP, NetBIOS, DECnet, and Banyan Vines network protocols. What this means is that a network manager can configure specific applications to use the remote access method that's best suited for optimum performance.

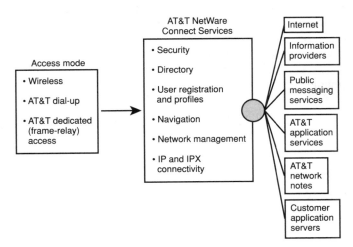

Figure 10.6 AT&T NetWare Connect Services.

The eight modems attach directly to the server's bus. The advantage of such an architecture is that serial port bottlenecks in the server are eliminated since data travels through the server on a packet-by-packet rather than byte-by-byte basis. The result is less overhead and transmission delays. Among the modem standards supported are V.fast (28.8 bps), V.32, V.32bis, V22, and V.22bis. The MNP class-10 error-correction protocol is also supported.

Network management of remote access servers is crucial for enterprise network managers. The Micom LANexpress server supports simple network management protocol (SNMP). While any third-party SNMP-based network management program can be used in conjunction with this server, Micom's expressWATCH provides real-time access to the remote server's components. A network manager can troubleshoot any component, including the modems, while they're in use.

Security on the LANexpress server takes the form of multiple security levels for dial-in and dial-out users. Each server keeps an encrypted user database to track individual user profiles and an event log to track user activity. Dial-in security requires a user name and password before connecting through the server. This password is encrypted, and each user profile can be configured for pass-through, fixed dial-back, or roving dial-back security.

The dial-out security consists of preventing unauthorized use of long-distance telephone services by verifying user name and password dial-out access rights before connecting the user to a modem. Each user can have either unlimited dial-out access or a fixed telephone number for limited access.

Each server keeps a security database of all users. The network manager (via password) can view, edit, or archive the encrypted database. Perhaps the most interesting security measure available from Micom comes as a result of the company's integration of security software from Security Dynamics into its server. This software is similar to that found in Security Dynamics' ACE/Server LAN. This particular server provides security via a credit-card-sized smartcard that displays a randomly generated, unpredictable access code that automatically changes every 60 seconds. To access a protected network, a user simply enters the secret memorized identification number followed by the access code appearing on the card.

There are dozens of these bundled remote access server products on the market. How do you tell one from the other? The following are features to look for:

- Proprietary compression algorithms to increase bandwidth optimization
- Diagnostic software to test the server and attached modems
- Password encryption
- Aging access-time restriction

- Multidestination dial-backs
- Protocols supported
- Length of warranty
- Number of modems and ports supported

The remote bridge

Imagine that a company is running NetWare on its corporate LAN and also on the small LAN found at its branch office. The branch office users don't need to be connected to the corporate LAN all the time, but they do need occasional access. The amount of bandwidth they need varies depending on what files they need to transmit and receive. In this particular situation, there's only one protocol (NetWare's IPX) involved, so no router is required. A remote bridge can connect the two LANs.

Combinet's EveryWare product line includes a remote bridge capable of handling the remote access needs of a branch office LAN's workgroup, supporting Ethernet's 10BaseT and 10Base2 configuration. Figure 10.7 demonstrates the way such a system would be configured. This particular product is capable of using either Switched-56 or ISDN.

There are a number of advantages to remote bridges. They're easy to configure and administer, and use common addressing schemes. Perhaps the major advantage is that they're very simple to install.

Boundary routing

Whether the term is *boundary router* or *access router* or even *periphery router*, the purpose is the same. This type of router has been designed specifically for remote offices that must communicate with other locations. While their features differ depending on the individual vendor, there are certain generalizations that apply to this whole family of routers. The price is generally low because many companies have to purchase several of these products with identical configuration for dozens or even hundreds of branch offices. Perhaps as a result of the effort to keep prices low, vendors offer limited wide area network interface options. Products can generally be configured for T-1 lines or ISDN.

Most boundary routers can have downloaded configurations. The router's operating system software usually resides in electrically erasable programmable ROM, or flash EPROM. Because these routers are located at branch offices where data traffic isn't particularly heavy and because WAN traffic is usually defined by the restraints imposed by limited bandwidth availability, there's limited throughput. Finally, protocol support is generally limited. There are exceptions, of course. Both Cisco and Wellfleet offer support for all protocols supported by their larger routers.

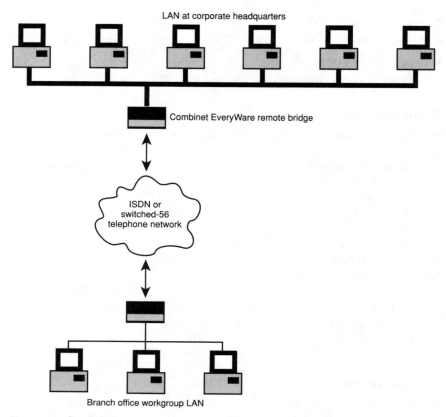

Figure 10.7 Combinet's remote bridge linking a branch office and corporate headquarters.

3Com has developed the concept of a boundary router, the premise of which is that routing intelligence and decision-making can be centralized to improve ease of installation. Its boundary routers decide whether to forward a packet but don't need to decide which WAN port to route packets to because there's only one port in operation at any time. A second port is used only for backup in the event of link failure.

Configuration and other network management functions are performed at the central site. Consequently, 3Com claims that boundary routers typically take less than 15 minutes to set up and configure versus hours for conventional routers, a fact that has been verified through independent testing.

Boundary routers essentially act like a bridge, passing all traffic across the remote link. New smart filters in 3Com's router reduce the unecessary NetWare IPX broadcast updates and service advertising protocol updates coming from the central router across the WAN by up to 90 percent, according to 3Com. These smart filters also enable the central site to learn and act on

TABLE 10.1 Responsibilities of the Boundary Router and Central Router in a 3Com Environment

Central router	Boundary router
Handles all routing decisions	Filters and forwards packets based on Ethernet addresses
Maintains routing tables	Provides customizable filters to reduce traffic crossing the wide area network
Provides all configuration information	Gathers SNMP statistics

traffic patterns from the remote site. Table 10.1 illustrates the responsibilities of the boundary router and of the central router in a 3Com environment, while Figure 10.8 shows a complex enterprise network environment where boundary routers coexist with centralized routers, LANs, and mainframe computers.

The intelligent hub

Many larger companies have already made considerable investments in their intelligent hubs; rather than introduce a router vendor's product into this environment and possibly create interoperability conflicts, they prefer looking to the intelligent hub vendor for a branch office product. Proteon was one of the first hub vendors to develop products specifically earmarked for the branch office environment. Its Branch Office Solutions (BOSS) family of hubs contain routers that support major network protocols, including NetWare's IPX/SPX, TCP/IP, and over 17 other protocols. Point-to-point protocol is supported, as well as frame relay and X.25. These hubs consolidate 3270 traffic with token-ring LANs over a single WAN connection, so it's economical for branch offices that need to access corporate mainframes as well as corporate LANs; both kinds of data can travel over the same wide area network links. Other intelligent hub vendors, such as market-leader SynOptics, also offer intelligent hubs with built-in router modules for branch office operation.

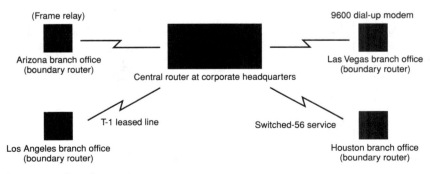

Figure 10.8 Boundary routers.

SynOptics' solution illustrates how various enterprise network components, such as backbone routers, branch office hubs with router modules,
and network management software, can be combined to offer a comprehensive solution to the problems a network manager faces in ensuring interoperability throughout an enterprise network. SynOptics' LattisEngine/486 is
an intelligent communications sever that integrates distributed communications services, such as multiprotocol routing, SNA gateway connectivity,
and system-level management into a hub architecture.

LattisEngine/486 is a routing/communications module that occupies two
slots in a SynOptics intelligent hub. As an alternative, network managers
with very small branch offices might opt for a stand-alone version of the
LattisEngine/486 that will work in conjunction with SynOptics' workgroup
hubs.

The routing/communications module, when configured with Novell's Multiprotocol Router Plus software, provides multiprotocol routing services for
NetWare LANs linked locally and over wide area networks. The module also
supports NetWare LAN-to-mainframe connectivity over IBM's system network architecture (SNA) networks when Novell's NetWare for SAA and
NetWare SNA links software is loaded. Figure 10.9 illustrates SynOptics'
view of an enterprise network environment that includes branch offices.

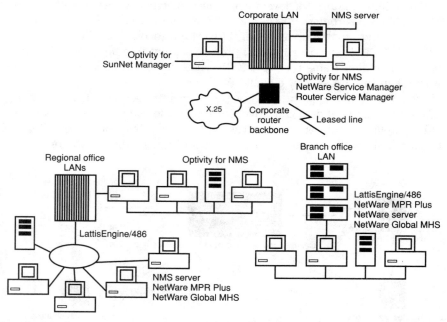

Figure 10.9 SynOptics' view of an enterprise network with branch offices linked to corporate
headquarters.

Note that the corporate LAN contains a network manager's console running Optivity for SunNet Manager software as well as Optivity for Novell's Network Management System (NMS) software. The end result is that a network manager at corporate headquarters can manage the entire enterprise network.

A hub in a router

Companies such as Retix and Madge Networks have developed a new type of product: a hub in a router. Retix's RX7000 family of products offer routers with WAN ports and also an optional 10BaseT 12-port card that fits into a single I/O slot within the router. A network manager who adds this unit to a branch office gains routing capability as well as the network management functions included with the hub card. Madge Networks offers a token-ring hub card that can be installed in a PC running Novell's Multiprotocol Router Plus software. The result is a very inexpensive branch office remote access server solution.

Factors that Could Enhance Branch-Office Connectivity

Several factors that could enhance branch-office connectivity are the speed with which an integrated services digital network (ISDN), point-to-point protocol, and IPXWAN protocol are supported by vendors of branch office products.

ISDN

Many branch office product vendors have been quick to offer ISDN, but the real issue is how quickly the regional Bell operating companies (RBOCs) implement this service throughout the country. It doesn't really help a company if half its branch offices are in areas where ISDN is available. All branch offices must have access to ISDN for a company to benefit from the available cost savings.

The vast majority (82 percent) of the approximately 3,300 U.S. establishments that installed ISDN by mid-1994 have more than 1,000 employees. If ISDN's limited strength is in large establishments, it's discouraging to realize that less than one percent of establishments with more than 1,000 employees planned to install ISDN before the end of 1994. The deployment schedule for ISDN is hopelessly optimistic, given past performance. Network managers must really investigate the availability of ISDN and not rely on telephone carriers' promises when developing a plan to link branch offices into a wide area network.

PPP

The Internet Engineering Task Force (IETF) established specifications for a point-to-point protocol (PPP), designed in order to create a standard way for routers to set up a session, monitor the link between them, and then terminate the session. The set of specifications defines the format for framing the PPP packets as well as some additional functions, such as security and compression.

It makes perfect sense that a standard protocol for linking remote offices would make it easier for network managers to use products from multiple vendors without encountering interoperability problems. Unfortunately, this isn't the case. PPP is defined in ten different RFC documents. Vendors need support only a single RFC to claim that their products are PPP-compliant.

The PPP protocol does enable routers to negotiate with each other before data is transmitted. It's possible, though, for two PPP-compliant routers to negotiate, define the features each supports, and determine that they have nothing in common. These negotiations take place transparently to the human operator, who can only wait to see if communications can proceed.

In short, PPP is certainly an important step toward remote communications interoperability; unfortunately, it still leaves too much chance that products won't be able to work together. To avoid this situation, consider buying routers from a single vendor.

IPXWAN

Because so many local and remote LANs run Novell NetWare and its IPX/SPX protocol, it's important to understand the benefits gained if a product supports IPXWAN. This extension to the Novell IPX protocol provides for the negotiation of the parameters required for communication over wide area network links. IPXWAN has been designed to operate over multiple WAN technologies, such as frame relay, ISDN, PPP, and X.25. Up until the release of this protocol extension, router vendors used different proprietary IPX negotiation methods for establishing and maintaining links between routers over a wide area network.

Because IPXWAN provides a standard way to ensure compatible configurations regardless of the vendor's product, it's a major step toward establishing interoperability among router products supporting IPX. Novell has promised additional negotiated features, including the ability to determine whether a link is to a workstation and whether a link is dynamic.

Questions to Ask When Considering Branch Office Connectivity Products

How difficult is it to upgrade branch office software? Because of the difficulty of personally visiting each branch office, network managers want branch office software that's easy to upgrade remotely. 3Com's boundary routers offer an example of a feature known as *dual flash*, which enables an old version of software to run while a new version is loaded. When the load is complete, the router automatically switches to the new version. If a problem exists, the router switches back to the old version of software until the problem is resolved.

How easy is it to install the branch office router and then make subsequent changes in configuration? Some router companies offer free network management utilities running on UNIX machines that permit a network manager to examine and change a router parameter. Other companies, such as Cisco, offer such software but at considerable expense. Look for software that provides a menu-driven installation rather than requiring that cryptic commands be typed at a command line. Are the protocols you need to send over a wide area network supported?

The remote routers on the market differ widely in the support of protocols, particularly some of the less common ones. DECnet and AppleTalk are supported by only a relatively few branch office routers. Cisco and Wellfleet tend to offer more complete protocol options than some of their competitors.

If throughput is a major consideration, what network interfaces and compression schemes are supported? If all branch offices have access to ISDN, obviously the remote routers should also support this type of transmission. Similarly, Advanced Computer Communications (ACC) has a set of features to optimize bandwidth across wide area networks. Its routers compress data four times faster than products offered by most of its competitors. Its express queuing feature uses an algorithm that determines transmission based on the order of packets ready for transmission; it also permits prioritization of packets.

Are the products fully configured? Cisco, for example, offers its entry-level branch office router at $1,000 below the price of Advanced Computer Communications' product. Yet, when both products are configured with similar software, the ACC product costs $1,000 less because it comes with software while the Cisco product requires a $2,300 software option in order to be comparably configured. Some vendors charge as much as $200 for a cable to connect a router to a modem.

Summary

Remote communications involve dial-up communication links in many cases, and dedicated leased lines, such as T-1 or T-3 links, when there's sufficient traffic to warrant them. Most PC LAN communications involve data coded in ASCII format and sent over lines in an asynchronous (stop/start) transmission. Error checking is particularly important for remote communications because it lacks the reliability found on local area networks. A number of error-checking schemes are available in conjunction with remote communications software and hardware, ranging from simple parity checking to complex cyclical redundancy checks (CRCs) that require elaborate mathematical algorithms.

Network managers can achieve high rates of data transmission using modems that support data compression. Many of these compression schemes are still proprietary, with Microm Network Protocol (MNP) providing the widest level of support. MNP class 10 provides support for international standards, as well as Microcom's proprietary standards.

There are two major ways to provide remote access to a LAN from remote locations such as branch offices. Using remote control, a remote user can seize control of a distant PC that's attached to a LAN and then use that PC to process applications. Only keystrokes and screen images are transmitted over the remote link. This approach is useful for limited access, such as e-mail, but doesn't provide full network access as does the remote node approach. The remote node approach uses a remote access server to make network linkage transparent to end users.

Remote access servers can take the form of a PC running router software and remote access software, bundled hardware and software products that include integrated modems, stand-alone routers, hubs that contain integrated routers and WAN interfaces, and routers that contain hub cards and WAN interfaces.

Among the factors that will increase the growth of branch office links to remote LANs is the growth of integrated services digital network (ISDN), point-to-point protocol, and IPXWAN protocol.

11

Network Management

In this chapter, you'll examine:

- How networks use simple network management protocol (SNMP) for management purposes
- Why SNMP has become so popular for network management
- The major network management approaches taken by IBM, AT&T, and DEC
- Desktop network management systems and the desktop management interface (DMI)

As networks evolve from single-site LANs to enterprise networks that include multiple LANs as well as remote LANs with WAN links to mainframes and minicomputers, it becomes even more crucial to be able to monitor all LAN components in a timely manner. In this chapter, you'll look at managing network desktops as well as larger systems.

Network managers and systems integrators grapple with the problem of integrating noncompatible, proprietary networks into an enterprise network. While planning this enterprise network, however, it's absolutely crucial to determine some way to monitor, manage, and control traffic over it. The following sections describe a network management protocol (SNMP) associated with TCP/IP networks, and another protocol (CMIP) associated with OSI networks.

Simple Network Management Protocol (SNMP)

For several years, network managers have been waiting for the ISO to finalize a set of network management protocols. By 1987 many industry insiders had become very impatient as the ISO grappled with a document that had grown to several thousand pages and seemed nowhere near completion. A group consisting of two professors from MIT, two professors from the University of Tennessee, and two engineers from Nysernet created simple network management protocol (SNMP). Designed originally as the network management protocol for TCP/IP, this protocol is now able to monitor network traffic and indicate both malfunctioning equipment and performance bottlenecks on a variety of non-TCP/IP network devices, including 802.1 Ethernet bridges.

The Internet Activities Board that governs TCP/IP has given standard status to several elements of SNMP, including the management information base I (MIB-I) and structure of management information (SMI). Today, over 80 vendors offer SNMP products and work is progressing on extending SNMP so it can handle 802.5 LANs and DECnet phase IV devices. SNMP was developed as a connectionless protocol to reduce overhead requirements and maintain user control rather than have parameters handled transparently as they are in a connection-oriented protocol.

The SNMP protocol links a network management system (NMS) and a device that's being managed. This device contains an agent that communicates with the NMS. Information is stored in a management information base (MIB). This database contains network statistics (packets transmitted, errors, etc.).

Since the MIB is defined in standard terms for managing an internet router, vendors must extend the MIB so information is relevant for their particular device. The NMS queries agents on a regular basis and receives responses. Under SNMP, agents are very small programs because their function is simply to respond to queries with only five different types of messages. The majority of the processing required is done at the NMS. This arrangement encourages vendors to write the required agent software for their products so they'll work in an SNMP environment.

For devices that use proprietary protocols and aren't yet SNMP-compatible, it's possible to use a proxy agent. A *proxy agent* is software that translates a proprietary protocol into SNMP and vice versa. The proxy agent receives information from an agent that uses the proprietary protocol and translates it into SNMP format before forwarding it to the NMS. While SNMP can't prevent something from happening directly, it can alert a manager when traffic patterns and conditions indicate that a device is about to go bad. A network manager can set thresholds for devices ("above x number of packets retransmitted," for example) and then receive an alarm when the threshold is reached.

SNMP is a transaction-oriented protocol that permits network managers to select any particular event they want to have a network agent initiate activity. By initiating an SNMP GET request, a user can view values of objects in the management information base (MIB).

Using SNMP to monitor a network

A network manager could monitor a network from a network management station (NMS), and the NMS would monitor and control SNMP agents found within network devices. Network displays are usually in real time, so managers can observe hanging conditions. Usually a crucial alert is shown in red. Managers can use an SNMPSTAT command to interrogate the variables associated with a particular device found in the SNMP database known as the management information base (MIB). The agents associated with each hub, router, and other managed devices take the requested MIB information and deliver into the NMS.

SNMP version 2

Industry concern over the lack of security found under SNMP resulted in the 1991 formation of a committee to develop a "secure SNMP." The committee soon decided to incorporate additional features into SNMP version 2, and developed a specification known as SNMPv2. This specification consists of 12 working documents totaling more than 400 pages.

This protocol offers four different levels of security. The first level, node security, is the same type of security—no authentication nor privacy—found under SNMPv1. The second level consists of verifying the identity of the network user using a protocol called message digest 5 (md5), which encrypts authentication information such as passwords. The third level is configuration privacy, which uses md5 but permits the network administrator to set up configuration files from anywhere on the network and encrypt the data with the federal information processing standard data encryption standard (DES). The final level offers full privacy. It uses md5 for authentication, DES for configuration, and encrypts all transmitted network management data with DES.

SNMPv2 contains a get bulk request protocol data unit, which allows a manager to retrieve a large amount of data with a single request. Under SNMPv1, each data request can generate only a single response. Another improvement offered by SNMPv2 is the inclusion of new data types and new conventions for managing the tables created by the structure of management information (SMI). The lack of manager-to-manager communications found under SNMPv1 has been corrected with the inclusion of M2M (manager-to-manager) MIB, which establishes the communication requirements between managers and submanagers.

There are some practical problems with SNMPv2. While it offers greater security as well as enhancements in data retrieval and manager-to-manager communications, it has some very serious limitations; the differences in SNMPv1 make the two protocols incompatible. The only temporary migration solution is to use network management stations capable of handling both protocols, though that's more overhead than most network managers want to consider. The complexity of SNMPv2 is making it difficult for vendors to implement. Security is very cumbersome and resource-intensive. Finally, the work required to implement SNMPv2 products will raise the price of SNMP management. Among the vendors committed to SNMPv2 are AT&T, Cabletron, SynOptics, and IBM. For the time being, network managers don't have a wide range of products to choose from that offer support for SNMPv2; it's still an SNMPv1 universe.

Common Management Information Protocol (CMIP)

Common management information protocol (CMIP) is a standard for network management over OSI networks. CMIP offers a much more robust set of tools for network management than SNMP. It provides six different types of services: configuration management, security management, fault management, accounting, performance management, and directory service.

A major distinction between CMIP and SNMP is that SNMP doesn't distinguish between an object and its attributes. An object might be a device and an attribute might be that device's condition or a parameter describing it. What this means in practical terms is that, while SNMP is easy to implement, it's difficult to maintain if a network is in a constant state of change. Under SNMP, a vendor would have to provide a different definition for each device it creates for an SNMP network, regardless of how similar these devices might be. Under CMIP, however, the vendor could let new devices use the same definitions used by other devices and simply include some additional attributes to distinguish them. The new products "inherit" the definitions of older products.

Another advantage of CMIP over SNMP is security. While work is progressing on security features for CMIP, SNMP has no security provisions. Unless network managers disable the SNMP SET command, anyone can write commands to SNMP devices and control them.

CMIP is slowly building up its own head of steam despite predictions that SNMP will dominate for several years. Digital's DECnet Phase V contains CMIP as its network management protocol. British Telecommunications PLC, MCI Communications Corp., and Telecom Canada have proposed to both the CCITT and to the OSI/Network Management Forum that CMIP be used over CCS7 (Signaling System 7). The advantage for end users is that these common carriers would be able to provide network management information concurrently with the data being transmitted. A gateway would provide the management data but lock users out of the internal operations of SS7.

Common management over TCP/IP (CMOT)

CMIP hasn't really taken off because of the small base of installed OSI networks. A network management group (NetMan group) composed of such industry giants as Hewlett-Packard, Sun Microsystems, Digital Equipment Corporation, and 3Com have come up with one solution to the major limitation to CMIP's growth—the lack of installed OSI networks. The common management over TCP/IP (CMOT) protocol consists of standard CMIP with a presentation layer that maps OSI layers 6 and below to TCP/IP.

CMOT was developed to encourage companies currently using TCP/IP but planning to migrate eventually to OSI networks to make the network management shift now—secure in the knowledge that they won't have to change their network management software later when they make the transition to OSI.

Distributed Management Environment (DME)

The Open Software Foundation's (OSF) distributed management environment (DME) project is working to establish an industry-wide standard model for systems management across all computing environments. Its goal is to have DME serve as a common platform to access network management services such as SNMP and CMIP. For DME to really catch on, it must specify the distributed management of the same type as the tasks currently managed on large systems. Such tasks include data security, software distribution, software maintenance, license metering, mapping of network devices, troubleshooting, client management and interrogation, and backup and archive functions. Part of DME is implemented in version 4.0 of Hewlett-Packard's OpenView distributed management platform.

IBM and Heterogeneous
LAN Network Management with OSI Protocols

IBM has announced a method of dealing with the non-IBM LANs that need to be monitored on an enterprise network. IBM and 3Com have announced a joint effort to define a set of network management specifications for these mixed-media LANs. Known as the heterogenous LAN management (HLM) specifications, they will provide tools for network managers to monitor, control, and analyze network data.

HLM incorporates a subset of the OSI common management information protocol (CMIP) and the IEEE's 802.2 logical link control (LLC). Known as CMOL, this protocol will help companies migrating their networks to full

OSI model compliance. The two companies will also develop application programming interfaces (APIs) to enable third-party developers to produce additional network management software.

LAN Desktop Network Management Software

The vast majority of network management software found on LANs today consists of various utilities, usually bundled together and then offered as a LAN desktop management package. Each utility performs a specific task. Among the major tasks most often performed by this software are monitoring server performance, metering software, monitoring network traffic, displaying packet statistics, scanning for viruses, managing LAN software and hardware inventory, and managing print queue operations. Examples of this software are Frye Computer Systems' Utilities for Frye Networks, Intel's LANDesk, Synmantic's Norton Administrator for Networks, and Microcom's LANlord.

In addition to the features already discussed, there are several features worth looking for that might help differentiate one LAN desktop network management software package from another:

- Software distribution
- Custom report writer
- Physical network monitoring
- Centralized monitoring across multiple LANs
- Event managers that can define notification priorities
- Automated responses via e-mail, beeper, etc.
- Monitoring of security violations

Novell's NetWare Management System (NMS)

Because Novell's NetWare is found on over 65% of the installed base of PCs on LANs, any network management system must be able to provide information on NetWare servers and NetWare LAN traffic. Novell's NetWare management system is able to manage NetWare servers as well as devices and systems using simple network management protocol (SNMP). Novell supplies a set of agents for NetWare server management using its own proprietary management protocol. Eventually, server management under NMS will be handled via SNMP.

NMS provides distributed network analysis by continually monitoring the interaction between network devices for more effective troubleshooting; it provides statistics on the level of traffic, identifying the most active nodes, and performs detailed packet analyses. NMS handles address management by discovering, storing, and displaying all internetwork protocol exchange (IPX) and internet protocol (IP) numbers in use, along with other information, including the physical media type. It's important to note that NMS does support SNMP via an enabling MIB agent. The NMS console receives alarms from the NetWare LANalyzer agent and the NetWare management agent if either discovers duplicate addresses. Mapping is created by NMS's ability to discover all IPX devices and IP routers.

Several hub vendors, such as SynOptics and Cabletron, support management applications for NMS. SynOptics supports NMS via its Optivity product, while Cabletron provides support via its Spectrum product. Intel's LANDesk Manager is an example of a desktop management program that takes advantage of NMS information.

The Desktop Management Interface (DMI)

The Desktop Management Task Force (DMTF) was initially formed with Intel, Microsoft, Novell, SunConnect, SynOptics, and, a bit later, Digital Equipment Corporation (DEC), Hewlett-Packard, and IBM. The goal of this group has been to develop a set of specifications for desktop management, a common interface to which developers can write management applications. Because manufacturers can write to a single interface, they don't have to understand which management protocol or application will be used to manage the component. They can provide the level of management they deem appropriate for their components.

In August of 1992, the DMTF announced an open API architecture for desktop management known as the desktop management interface (DMI). This set of specifications includes a management interface (MI), enables DMI-compliant applications to access and control desktop PCs and their components. The DMTF has also announced a simple component interface (CI), which can be implemented by software and hardware component vendors in existing and future products. The CI allows products to be managed by applications calling the DMI. As a result, component vendors are shielded from decisions about management applications and protocols, and can focus on providing competitive management of their products.

The DMTF is developing DMI test implementations at the management interface level for SNMP and CMOL. It has also indicated an interest in extending DMI to platforms other than those that are Intel-based.

Novell's NetWare Distributed Management Services (NDMS)

Novell has announced that its NetWare distributed management services (NDMS) strategy is to link vital network services such as software distribu-

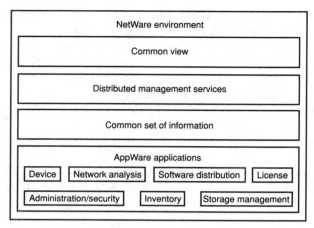

Figure 11.1   The structure of NetWare's distributed management services.

tion, network administration, device management, storage management, inventory analysis, and network analysis and control to its NetWare Management System (NMS). Figure 11.1 reveals the structure of NDMS.

Part of what Novell needs in order to develop NDMS is already in place. The company has its NMS console, AppWare development platform, Navigator software-distribution product, and LANalyzer agent multisegment network analysis technology. AppWare is Novell's development package that enables developers to produce software that can be moved from platform to platform under NetWare. The NetWare Navigator is a set of NetWare loadable modules (NLMs) that enable network managers to distribute and install application and operating system software automatically. The LANalyzer agent is a set of NLMs that monitors traffic on network segments and sends alarms to the NMS operator console. The NetWare licensing system software development kit (SDK) enables network software developers to write electronic software licensing and related software tracking and management applications that complement NDMS software licensing services.

Novell also offers a NetWare hub services agent, which monitors and manages hubs that comply with the Novell hub management interface specification. The result is a series of graphical views of hubs that enable a network manager to enable or disable ports from a central management console. Novell's plan is to manage devices using NMS and SNMP, analyze the network using LANalyzer agent technology, manage software distribution and licensing using NetWare Navigator, and manage storage via storage management services (NMS).

Novell plans a standards-based approach rather than using proprietary protocols. In fact, the company wants the openness of its platform to make information accessible by third-party applications as well as by other platforms such as SunNet Manager and OpenView. Under NDMS, NetWare servers would feed a common distributed database of information on users, network directories, applications, etc. This distributed information is referred to by Novell as *management services*.

Novell's goal is that a network manager will be able to use a graphical user interface (an NMS console) to manage all the network services previously listed. All these services will share a common database and even management system. Novell has worked with Intel, so its LANDesk Manager is integrated closely with NMS.

One major disappointment for many network managers who have both UNIX and NetWare as part of their enterprise networks is Novell's decision to discontinue development of its distributed manager UNIX platform (DM). Distributed manager provided management of users and groups, devices, software licensing, software management, systems management, and management of backup and recovery. Novell introduced DM in UNIXWare as a rival to distributed management environment (DME). The company has promised that much of the DM technologies will appear eventually in NDMS.

Microsoft's Hermes

Microsoft has made no secret of its desire to offer superior desktop network management via its Hermes database, which competes directly with Novell's NDMS. A major problem faced by companies that are running both Windows NT and NetWare is that, while both Hermes and NDMS support SNMP, they don't support each other. Digital, Legent, Compaq, Hewlett-Packard, SynOptics, and 18 other companies have announced that they will develop hooks to their software for Hermes. Hewlett-Packard and SynOptics have announced that they'll also support Novell's NDMS. Other companies supporting NDMS include Borland, Lotus, and WordPerfect. Of course, WordPerfect is owned by Novell, so its support comes as no surprise. Fortune 500 MIS directors are likely to apply pressure to force Novell and Microsoft to bury their differences and offer a level of interoperability. The implied threat is that, unless this is done, these companies can take their business elsewhere, perhaps to UNIX.

When released, Hermes will be known as System Management Server (SMS). This product will contain software distribution, software metering, hardware and software inventory, and network monitoring modules. SMS's automatic software upgrades will take into account system configurations and software version numbers.

SMS will require a terminate-and-stay-resident (TSR) program on clients to allow automatic software updates and remote troubleshooting. Client software will be available for MS-DOS, Windows, Windows for Workgroups, and Windows NT, as well as the Macintosh. SMS is capable of automatically propagating its client software over the network to any machine that doesn't already have it.

Hub-Based Network Management Software

Hub vendors have taken the lead in developing network management software that not only manages their own intelligent hubs but also provides this key information to popular network management platforms such as Novell's NMS, SunConnect's SunNet Manager, Hewlett-Packard's OpenView, and IBM's NetView/6000. Cabletron's Spectrum data gateway enables the Spectrum network management software to exchange data with the popular network management platforms mentioned previously. The latest release of Spectrum embraces the concept of distributed network management across an enterprise, and includes the first fruits of artificial intelligence technology that both Cabletron and its rival SynOptics have been pursuing. The product is designed to increase the ability to automate many network management functions. It offers the ability to link multiple management servers, known as "spectro servers," to create a single database. The purpose of this distribution of management functions is to provide the flexibility for local, regional, and even global management.

SynOptics has also been working to distribute its network management functions throughout a network. It has developed an architecture known as global enterprise management (GEM) to enable vendors to develop applications to take advantage of the embedded intelligence distributed throughout the network fabric. A single, high-level application will be able to manage multiple technologies and products from multiple vendors.

SynOptics has developed "superagents"—sophisticated network management software that can be distributed throughout an enterprise network to perform specific management tasks and feed the information to GEM applications as well as to vertical solutions, such as SynOptics' Optivity software. These superagents coordinate the activities of lower-level device agents, consolidate this information, and pass it to higher-level management applications. Superagents are specifically designed for such functions as local fault correlation, inventory auditing, and topology mapping. Their major role is to relieve the central management station of low-level, time-consuming processing chores. SynOptics is working with Intel, Hewlett-Packard, IBM, Novell, and SunConnect to develop GEM applications. Figure 11.2 demonstrates how these superagents function in a GEM environment.

Network Management Systems (NMSs)

A network management system (NMS) manages and controls complex networks, and is able to reconfigure the network when workstations and other resources are added or deleted. It can filter the thousands of alerts it re-

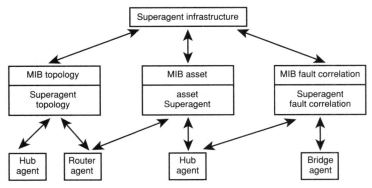

Figure 11.2 Superagents in SynOptics' GEM environment.

ceives and reduce the "noise level" down to where a network manager can view an uncluttered screen and see the most significant network events taking place. The NMS is able to manage bandwidth, provide cost accounting and performance measurement, and monitor security all in a user-friendly graphics environment. The NMS described in this paragraph is the ideal; all the major vendors are still far from fulfilling this order, but they do have clear visions of their own ultimate NMS, as well as actual products available today.

The major theme in this book is linking LANs to form enterprise networks that include wide area networks. While the LAN desktop management software described in this chapter is adequate for stand-alone LANs, it is clearly not adequate to meet the needs of an enterprise LAN. The answer is a LAN network management system that can accept the information fed to it from devices that include LAN desktop PCs as well as routers, hubs, etc., on dozens of individual LAN segments. Let's examine some of the major network management systems available to the network manager.

Hewlett-Packard's OpenView

Hewlett-Packard's OpenView is a management platform that can integrate network information from NetWare LANs and UNIX networks. This platform supports SNMP, CMOT, and CMIP protocols and provides a migration path to the Open Software Foundation's DME. A Windows version provides concurrent use of TCP/IP and IPX protocols. OpenView currently holds the leading share of market for management of UNIX networks. OpenView provides automated filtering, customizing, and extending of events and alarms. It also has the ability to discover devices, map them, and then display them in a variety of colors so that alerts, for example, can be depicted in red for immediate attention. Hewlett-Packard has been working with third-party vendors to bring management information from DECnet, Systems Network

Architecture (SNA), and NetWare into the OpenView database, which is based on the Ingres database engine.

The next version of OpenView, currently code-named Tornado, will be able to integrate a variety of applications across a management network while consolidating management data in a central repository. Local management functions will include data collection, event monitoring and reporting, status polling, and network discovery.

This new release will permit users to link multiple OpenView workstations in a distributed management network that shares a common repository of management data. This type of architecture will make OpenView capable of managing enterprise networks with thousands of nodes. The product will support SNMPv2 and thus provide a higher level of security and distributability. The common repository will permit network administrators to view various network types, such as DECnet, SNA, and NetWare, on a single map.

SunConnect's SunNet Manager

SunConnect's SunNet Manager is a distributed network management platform for the Solaris multitasking operating system. It supports SNMPv2 and DMTF's DMI and DME standards. SunConnect has begun the task of integrating NetLab's DiMONs 3G object-oriented technology into its platform, which will enable multiple network managers to manage a network from any station on a LAN. SunNet Manager supports integrated tools for fault, configuration, performance, accounting, and security management services. The result will be Encompass, a fully distributed, object-oriented architecture. All the services provided by the DiMONs technology, including event management, communications, and database, can be centralized to create a distributed environment in which applications can run on separate computers and yet still share common services and data.

A major breakthrough for SunConnect came with its alliance with Novell and its relationship between SunNet Manager and NetWare Manage-ment System (NMS). Both companies agreed that SunNet Manager will provide central administrators with an overview of the enterprise network and let NMS users concentrate on specific NetWare issues. The advantage to network managers is that many have networks where NetWare and UNIX are coresident. They now can manage both UNIX boxes and LANs from a single console. SunNet Manager will access detailed information from NMS regarding NetWare servers, print queues, print jobs, LAN subsystems, NetWare loadable modules, memory, CPU and disk use, and software configurations. The companies have agreed to develop an autodiscovery tool that will allow a SunNet administrator to map out and diagram topology information about NetWare servers and attached devices.

IBM's LAN NetView

IBM's LAN NetView is the low end of a structure of IBM network management platforms that are designed to report upwards to the workstation and UNIX-based NetView/6000 and then to the mainframe NetView. Currently it can report only directly to the mainframe NetView. IBM's plan is to offer NetView/6000 as an enterprise manager for managing LAN NetView stations on distributed LANs, while the mainframe-based NetView serves as the enterprise manager of the entire network. IBM's LAN NetView Manage is a core network management program that, when combined with IBM's LAN NetView Monitor and LAN NetView Fix, is designed to provide distributed LAN management.

Lan NetView Manage's functions include discovering network resources, displaying graphical network topologies, linking management applications to LAN resources, and alerting network managers when problems occur. LAN NetView Fix focuses on event handling. Customers can develop software that can use the application to automate their problem determination procedures. LAN NetView Monitor is concerned with automated performance management. A network manager can use this software to specify the type of information to be collected, schedules for collecting information, thresholds to be set, etc. Monitor can also provide performance reports.

By adding appropriate applications from IBM or third-party developers, a network manager can have centralized or remote network management in a distributed environment.

The problem is that today's LAN NetView is quite limited. It can identify physical network problems, such as cabling problems. It has no special tools for managing a LAN server as a network operating system. It can track software and peripherals on NetWare servers and is able to determine what software is running on PC clients and OS/2-based file servers. The usefulness of this product will depend on what third-party applications are developed to support it.

IBM's NetView

IBM's view of network management is built around its mainframe computers and a traditional view of centralized processing. The NetView program running on a mainframe serves as a *focal point*, which receives information from entry points on various SNA NAUs. NetView release 2 provides an alternative LU 6.2 focal point to the mainframe program for non-SNA devices. This interface permits programmers to use APIs to create this link so that NetView can monitor non-IBM equipment.

NetView is able to communicate with LANs via key network management vector transport (NMVT), protocol-based code points. These alert code points are numbers that represent specific statements that appear on a

NetView screen. NetView's version 2 includes an automation table containing a set of commands that can be invoked when alarms, alerts, and other messages are received from a network device.

According to several recent industry reports, IBM and Novell have been cooperating so that NetView's version 2 release 2 will contain several NetWare-specific NMVT code points. IBM is expected to work with other vendors so they'll be able to develop code points for their networks to also be managed by NetView.

Non-IBM devices can communicate with NetView via NetView/PC, which runs on a PC and serves as a *service point*, IBM's definition of an interface to NetView for non-IBM equipment. NetView/PC uses a systems services control point to physical unit (SSCP-to-PU) session with the VTAM program running on the mainframe to pass control information to the NetView program. A token-ring network running LAN Manager can transmit performance and alert data to NetView/PC, which in turn forwards the information to NetView on the host via VTAM.

NetView can also accept voice information, particularly billing information. As Figure 11.3 illustrates, a call detail collector transmits data to the PC, which forwards them to the mainframe. The NetView/PC Rolm alert monitor sends alerts from the Rolm PBX to the NetView program running on the host.

NetView now also supports TCP/IP. IBM's TCP/IP version 2 for VM can transmit SNMP data from devices attached to an IBM host to a TCP/IP net management system.

While NetView's major strength is its ability to monitor SNA networks and IBM's own equipment, it does have a number of limitations. Companies with non-Rolm telephone equipment, such as an AT&T or Northern Telecom

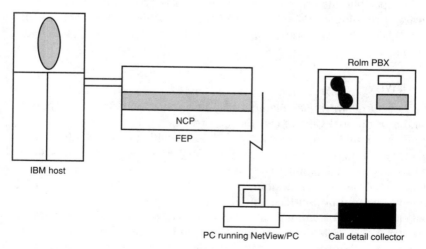

Figure 11.3 NetView collecting voice and data information.

PBX, have not been able to monitor their voice networks through NetView. Similarly, NetView can't communicate with non-IBM controllers or with LANs other than IBM's own Token Ring.

IBM's LAN Manager can be linked directly with NetView running on a mainframe. The product provides NetView with alerts, LAN status, and information on bridge and adapter functions on a multi-LAN bridged network. LAN Manager enables a network manager to check on a specific workstation's adapter card or bridge as well as actually change the configuration of a network segment or a bridge.

IBM's SystemView

In September 1990, IBM announced a systems architecture strategy that will eventually give network managers a consistent end-user interface (the end-user dimension), as well as the ability to share network management information across the entire IBM product line. At the heart of SystemView is a central database, a repository that will hold key systems management data collected across an enterprise network. IBM envisions SystemView as an open architecture that will eventually integrate SNA, SAA, OSI, TCP/IP, and AIX (IBM's UNIX) networks.

Why SystemView? Some industry experts see it as a typical IBM ploy. Announce a new product to be delivered in the distant future to prevent customers from buying rival products now. IBM's SAA incorporates IBM mainframe networks running under MVS and VM with AS/400 running under OS/400 and PCs running under OS/2. What about UNIX? SystemView promises to bring networks running under IBM's own version of UNIX into this enterprise network.

SystemView's Repository Manager will run on the DB2 database. This database manager will have to be able to accept data from other non-IBM databases in order to provide meaningful network management and control.

IBM's SystemView announcement was far more visionary than specific. SystemView is a plan, a network architecture for the future that will include a common user interface, a common database, standard management data definitions that can be adapted by third-party vendors, and modular management applications. IBM's announcements that its future networks will use the OSI model's CMIP protocols and X.400, and also understand TCP/IP, suggests the broad scope of SystemView.

Winston Churchill once wrote a history of the world that included detailed descriptions of King Arthur and Camelot. When confronted by a critic who questioned whether Camelot and King Arthur ever existed, Churchill replied that if they hadn't, they should have. We can say the same thing about SystemView; it's a necessity for large networks that need to tie together their diverse elements. If SystemView doesn't exist, it should.

AT&T's network management

Given its data processing roots, it's no surprise that IBM chose to approach network management from a mainframe perspective. It's also no surprise that AT&T took a completely different approach toward network management, one based on its roots in voice communications.

Unified network management architecture (UNMA) stresses the voice side of network management and takes a distributed rather than centralized approach. Just as the nation's phone system is built on the foundation of standard protocols, UNMA is based on the OSI model, but incorporates a network interface known as network management protocol (NMP), whose specifications are available to third-party vendors. NMP closely resembles the OSI suite of network management protocols.

AT&T looks at network management data coming from three distinctly different sources: customer premises, the local exchange carrier, and the interexchange carrier. Notice how this orientation encompasses both voice and data information being transmitted over phone lines. The customer premises might include a mainframe, minicomputers, and LANs, along with a private branch exchange (PBX) phone system. Each of these network components might be managed by an element management system (EMS) designed specifically for that type of equipment. Unfortunately, these EMSs are device-specific and usually unable to manage an entire network.

As Figure 11.4 illustrates, various EMSs are linked together via network management protocols into a network management system that contains function-specific modules responsible for integrating data from different EMSs. AT&T froze the OSI model's CMIP protocols during its development and then published these specifications for other vendors' use. AT&T hopes that other vendors will develop EMSs that follow this network management protocol.

AT&T expects the functionality of UNMA to grow as vendors develop additional products. AT&T has indicated that there are nine major network

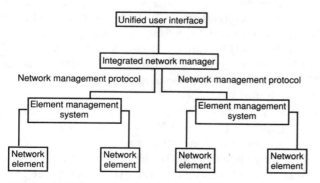

Figure 11.4 Unified network management architecture.

management functions that its system should be able to handle: configuration management, fault management, performance management, accounting management, security management, planning capability, operations support, programmability, and integrated control.

As the originator of UNIX, AT&T has predictably chosen to build its network management system around UNIX-based workstations or minicomputers running a program known as Accumaster Integrator.

Accumaster collects information from all major network components including T-1 lines, LANs, PBXs, and host applications. Information is filtered and then presented via a graphic interface. While Accumaster provides a single view of an entire network and allows you to pinpoint the source of alarms, it doesn't control all network elements from a single location the way NetView does. There are security and operational reasons why many companies with very large networks prefer this approach.

UNMA and Accumaster work hand in hand with NetView. Accumaster Integrator running on an AT&T 3B2/600 emulates an SNA PU 2.0 node. A Sun workstation can provide a graphic user interface, and a 3270 terminal emulation package provides the SNA connection via a 9,600-bps dedicated line.

With UNMA and Accumaster linked to an SNA network via NetView, an operator could observe a session alarm on the SNA network, use a mouse to select an option to graphically display the logical SNA session route mapped to the physical session route. After analyzing this graphic representation, the operator could identify the network problem and take corrective action. Figure 11.5 illustrates AT&T's UNMA with Accumaster Integrator in action.

UNMA is not yet a finished product. As more companies implement OSI protocols, AT&T's product will become more attractive because of its OSI-based network management protocols. As more enterprise networks develop, with an emphasis on long-distance lines and remote bridging, UNMA should become more popular because monitoring and controlling elements of a wide area network is AT&T's major strength.

What about UNMA and SNMP? After all, there are far more SNMP-based networks today than CIMP-based networks. AT&T has been working with third-party vendors to develop a network management system for NetWare file servers that will serve as a collection point for devices supporting SNMP. Alerts from SNMP devices would be translated and then transmitted via a gateway to Accumaster.

DEC's DECnet

Digital Equipment Corporation (DEC) released its first version of DECnet back in 1975 to enable two PDP-11 minicomputers to communicate with each other. The current version of this network architecture, DECnet Phase IV, was introduced in 1982. DECnet IV supports Ethernet as its LAN proto-

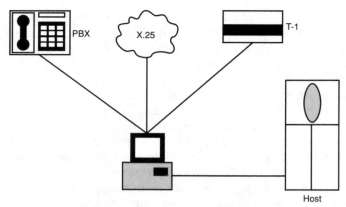

Figure 11.5 AT&T's UNMA and the Accumaster Integrator.

col, the X.25 protocol for packet-switched networks, and an SNA gateway, in which it functions as a PU type 2.0.

As shown in Figure 11.6, DECnet IV has eight layers, compared to the seven found in the OSI model. The DECnet IV physical layer corresponds closely with its OSI counterpart. The DECnet data link layer supports DEC's own digital data communications message protocol (DDCMP), the X.25 packet-switched protocol, and DEC's own Ethernet protocol, which isn't compatible with the IEEE 802.3 protocol used by the OSI model.

DECnet's routing layer uses proprietary DEC protocols to route data packets from node to node. The DECnet transport layer corresponds roughly with its OSI model counterpart but uses its own proprietary protocols. The session control layer performs the traditional OSI session layer functions of initiating and maintaining a session, but adds additional functions, such as translating DECnet node names to DECnet addresses and maintaining security by preventing unauthorized access to node resources.

DECnet's network application layer provides file transfer, virtual terminal service, and access to both SNA and X.25 gateways. It corresponds with the OSI model's presentation layer. The user and network management layers together provide the services offered by OSI's application layer.

DECnet Phase V

DECnet's Phase V has been a long time in coming. It adds the X.21 protocol for circuit-switched networks, and its data link layer adds both high-level data link control (HDLC) and OSI IEEE 802.3 protocols for Ethernet networks. Its network layer includes ISO CMIP protocols while still supporting management of Phase IV nodes. DECnet sheds its proprietary transport protocol and network routing protocols for ISO standard protocols. DEC's proprietary network information and control exchange (NICE) protocol for network management applications is replaced by the ISO's CMIP. DECnet V requires DEC's own naming service, VAX distributed name service (DNS), which was introduced back in 1987. DEC has committed itself to make this naming service compatible with the X.500 directory standard when it emerges in its final form.

DEC's enterprise management architecture (EMA)

DEC has quietly been building a platform for enterprise networking based on OSI model protocols. DEC has published the specifications for its enterprise management architecture (EMA) because vendors will have to incorporate the appropriate APIs in their network management software so these programs will be able to be a part of EMA. EMA defines management functions to include all five listed under the OSI model: configuration, fault, performance, security, and accounting.

Conceptually, the major components of DEC's enterprise management architecture consists of an "executive": access, functional, and presentation modules. The executive provides the operating environment, the open interfaces to non-DEC as well as DEC products, and the central repository for network management information.

Access modules provide the appropriate protocol to manage network entities such as computer systems, application programs, and modems. DECnet V entities are managed with the ISO's CMIP protocol. Functional

Application layer	User applications network management
Presentation layer	Network applications
Session layer	DECnet session control
Transport layer	DECnet transport
Network layer	DECnet routing (adaptive routing)
Data link layer	• Digital data communications messaging protocol • High-level data link control
Physical layer	Ethernet

Figure 11.6 DECnet IV and the OSI model.

modules are plug-in applications from DEC and other vendors that correspond to the five OSI management areas previously mentioned. The presentation modules integrate all EMA information through a common user interface. These modules could provide such different user interfaces as DEC windows, graphical user interfaces (GUIs), and ASCII terminals.

Unlike SystemView, which retains IBM's centralized network management and philosophy or UNMA with its strictly distributed management, enterprise management architecture is a platform that permits either centralized or distributed network control. The key to enterprise management architecture is the director kernel, which runs under DEC's proprietary VMS operating system and implements APIs that enable modules on different computers to interact without knowing the precise location of each other. Under EMA there can be one director that manages the entire network, several directors that manage the same network entity, or several directors that manage different network entities and communicate with each other. Developers who follow DEC guidelines can implement management modules that will interoperate under EMA.

The director is initially expected to support CMIP, SNMP, and DECnet Phase IV protocols. The director draws much of its data from a centralized network management database. The database, known as the management information repository, contains real-time performance and alarm status information. This database is consistent with the ISO's naming and addressing conventions, so its common structure for network element information can be accessed by non-DEC software.

Unlike IBM's SNA, in which nonintelligent devices are controlled, DEC has a tradition of building intelligence into its devices. Its bridges and DECnet nodes can make intelligent management decisions in a distributed environment. Since DEC is building EMA to incorporate its family of products, these devices are already equipped to send network management information to each other.

DEC released its first version of EMA as DECmcc in 1990. It included the DECmcc director, basic management system, site management station, and enterprise management station. The basic management system includes the director as well as DECnet—VAX software to communicate with VAX networks running under Ethernet or IEEE 802.3. The site management station includes the ability to monitor the network for alarms, calculate network performance statistics (LAN Traffic Monitor), store and forward historical management data, and receive notification of internal and external alarm conditions. The DEC extended LAN management software (DECelms) included with the site management station provides the ability to manage Ethernet/802.3 bridges and FDDI components in an extended LAN.

The enterprise management station includes all software found in the site management station, but adds WAN management using the NMCC/DECnet monitor for configuration, fault, and performance management of WANs. An optional TCP/IP SNMP access module links the EMA network to UNIX networks as well as others using SNMP management protocol. DEC has signed an agreement with System Center to jointly develop technology to allow DECmcc users to monitor and control SNA networks, and IBM NetView and Systems Center Net/Master users to control DEC systems.

One major weakness of EMA so far seems to be its lack of ability to link to NetWare LANs, a major component in many enterprise networks. Another problem is that most DEC LAN management tools, such as LAN Traffic Monitor, run outside of EMA although they can be combined with the director on a workstation called NetStation.

Summary

Simple network management protocol (SNMP) has become the de facto standard for managing local area networks. An enhanced version, SNMPv2, offers enhanced security and the ability to retrieve large amounts of data in a single request. The ISO's common management information protocol (CMIP) does offer greater functionality. A temporary measure for companies planning to migrate eventually to an OSI model compatibility is common management protocol over TCP/IP (CMOT).

IBM's approach to network management, NetView, is evolving. NetView now includes direct access from token-ring LANs and non-IBM equipment via LU 6.2. NetView can control and monitor Rolm telephone equipment (via NetView/PC) as well as SNA and token-ring networks. IBM's plans for network management (SystemView) incorporate UNIX-based computer systems and use standard ISO network management protocols (CMIP), while retaining the ability to use SNMP when needed.

AT&T's unified network management architecture (UNMA) uses a distributed approach. Accumaster Integrator workstations permit network managers to monitor network conditions from a centralized location, but control is distributed. DEC's enterprise network management architecture permits either centralized or distributed network management and control using a single or multiple directors. It adheres closely to ISO standards, including CMIP.

Hewlett-Packard's OpenView is one of the leading network management systems in the SNMP environment. Its next release will include the ability to manage distributed networks and view networks under such different systems as SNA, DECnet, and NetWare. The Desktop Management Task Force (DMTF) has developed a desktop management interface (DMI) that will make it far easier to manage desktop environments. Vendors will be able to write to this set of specifications and not have to develop their own APIs.

12

Major Systems Integration Issues

In this chapter, you'll examine:

- How fax servers can be incorporated in a network
- How networks can use the CCITT X.400 and X.500 recommendations to link proprietary, noncompatible e-mail systems via gateways
- Ways to make a complex enterprise network more secure

As networks evolve from single-site LANs to enterprise networks that include multiple LANs as well as remote LANs with WAN links to mainframes and minicomputers, integration of basic functions, such as e-mail, become more difficult and the need for centralized management and control increases dramatically. In this chapter, you'll examine how large multivendor networks can link heterogeneous e-mail packages. You'll also look at how large networks can share fax services. As networks become more complex, security is an even more crucial issue. You'll see how to make networks more secure.

Incorporating Faxes in an Enterprise Network

The fax explosion over the past few years means that LAN managers have needed to develop a cost-effective way for workstations to share fax services. Just having a fax available for several nodes to share isn't enough, however, because it's essential that the fax blend seamlessly with existing

LAN electronic mail software. When fax services are able to hide behind existing e-mail software, network users don't have to learn anything new before they can begin working efficiently.

Fax gateways exist for most major e-mail systems, which allow users to send faxes directly from their e-mail user interface. Such fax gateway programs usually contain interfaces to wide area network mail services, such as MCI Mail. Such programs are able to receive faxes and forward them to the appropriate user mailbox.

Fax gateways can be accessed by remote workstations. For example, using an enterprise network that happens to be using Higgins e-mail software with Higgins 2:Fax, a user on a remote LAN could send a message through several different "hops" until it reaches a network fax gateway.

Some points to consider about fax gateways

There are some negatives to adding a fax gateway to a LAN. Fax services place quite a strain on a PC's processing power. While fax operations can take place in background mode, they do slow up PC processing considerably as a network document is converted into the format required for fax transmission or a document is received and its contents must be converted into a standard ASCII text file. Does a network manager really want the network congested with the heavy traffic associated with fax transmission? Also, a fax server or gateway requires a scanner if the company needs to be able to send printed documents that aren't on the network.

Companies that use compression techniques to reduce 64-Kbps voice channels to 32 or 24 Kbps might not support 9.6-Kbps group 3 faxes. The multiplexers that compress voice information aren't designed to handle the variable-frequency tones used by analog fax transmissions. The fax might not be able to support 9.6-Kbps traffic and automatically seek a slower speed, such as 7.2 or 4.8 Kbps. There are some specific multiplexers, such as Newbridge's 3600, that direct fax traffic to a fax modem at speeds of 4.8 or 9.6 Kbps so the fax traffic can be transmitted along with compressed voice information.

LAN fax features worth considering

Selecting a LAN fax also presents problems. CCITT group 3 faxes are currently the norm; they can transmit a page of information in approximately one minute. Within the next few years, though, group 4 machines will become more common. Group 4 machines can transmit a page of information in around ten seconds. More important, though, this type of fax is designed to work with ISDN. Companies considering ISDN or involved in ISDN beta testing might want to look at a group 4 machine so they won't have compatibility problems in the future. Some models offer a group 3 converter so that the machines can communicate with the most common type of fax now in use.

Another thing for network managers to consider is the type of files the fax server can handle. Assuming a network manager has standardized the LAN on a specific word processing program or desktop publishing program, the fax server should be able to handle these types of files. If a LAN's user wants to send graphics files over a fax, then the fax server should be able to handle such common formats as Z-Soft's PCX, Digital Research's GEM, and Cybernetics' Dr. HALO. In the near future, more LAN users might need to send files in the Windows Paint format as well as the tagged-image file format (TIFF) used by many desktop publishing programs.

Still another feature that's essential for a fax gateway is network management. Comwave's Faxnet will serve as an example of some of the network management features a LAN manager can expect from such a product. Faxnet is a fax gateway designed to run on NetWare LANs. A DOS workstation can be configured with up to eight boards, with each board supporting a different fax line.

Faxnet can be programmed to prioritize outgoing faxes, store them, and then forward them at scheduled times. The server can be programmed to poll on-line services at scheduled times for specified data. Faxnet keeps track of each fax's destination, date, time, and duration. This information goes into a log as well as back to the source workstation.

FaxPress on a NetWare LAN

Let's look at a specific fax product designed to work with a NetWare LAN running under Ethernet, token ring, or Arcnet. Castelle's FaxPress isn't the only product available, but it will illustrate how this whole class of products operate.

As Figure 12.1 indicates, FaxPress is a dedicated fax server that enables a NetWare LAN to send and receive faxes from within any PC application. All fonts, graphics, and formatting are retained so the documents sent via fax look exactly as if they were printed on a laser printer. The server comes with either 2.5MB or 4.5MB of RAM and dual Motorola 68000 microprocessors. Standard CCITT group 3 faxes are transmitted at anywhere from 2,400 to 9,600 bps.

What's interesting about this new technology is that it seamlessly integrates all the advantages of a stand-alone fax machine into a NetWare environment. A NetWare user can send a fax four different ways: from within applications using a pop-up utility, from a menu, as an attachment to electronic mail messages, and from a command line. A user who needs a hard copy of a document and not a fax can use the printer attached to the fax server, so the fax server becomes just another NetWare print server when needed.

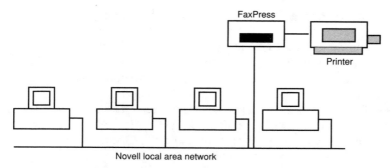

Figure 12.1 FaxPress as a NetWare print server.

The stand-alone fax can be programmed to send messages at specified times, even if the originating workstation isn't in use. Conversely, a fax arriving for a workstation that isn't logged onto the network can be printed. Network management features include three levels of password security and detailed logs of all fax transmissions.

Standards that improve fax transmission

A communications standard, V.17, modifies the V.33 communications standard for leased lines to work in the fax environment over conventional dial-up lines. The V.17 set of specifications increases transmission speed from 9,600 to 14,400 bps over current fax equipment by using a type of error correction (Trellis encoding) that's built right into the modulation scheme without any additional overhead.

The EIA 578 (asynchronous facsimile DCE control standard, service class 1) set of specifications defines a standard command set that can be incorporated on the fax board in a PC by a systems integrator. The specifications for class 1 mean the PC has to handle fax compression and connection while the fax board handles all other functions.

The major importance of EIA 578 is that it provides a standard set of fax drivers for systems integrators to use while writing software for programs running under Microsoft's Windows. Future classes of EIA 578 specifications will address such issues as binary file transfer and error control.

Problems with network fax gateways

If a fax server is used without a link to e-mail, one problem is that a networked fax won't know where to send an incoming file. Programs require a fax manager to scan incoming fax cover pages to decide where to route the messages. The EIA Fax Committee (TR29) is trying to develop a standard for routing incoming fax messages directly to their workstation destinations. The committee wants to develop a solution that will be backward compatible to the fax machines currently in the field.

What happens if an answering machine and a fax machine share the same line? Some fax machines will time out and then disconnect if they don't hear the appropriate fax tones. The CCITT Study Group VIII is working to develop standards for network fax servers that will be able to distinguish among voice, fax, and data modem calls.

Security presents still another problem when adding a fax gateway to a network. An enterprise network with several different fax gateways could present problems unless all gateways used the same proprietary encryption methods. At this time, the CCITT hasn't yet established standards for limiting access to networks with fax machines.

Managing Electronic Mail on Enterprise Networks

Electronic mail is probably the single most popular network feature for many people. As long as all company networks use the same e-mail package, there are no connectivity issues. For several years, TCP/IP's simple mail transfer protocol (SMTP) provided a "bare bones" common language that enabled users to exchange messages. Since TCP/IP wasn't available for PC workstations until recently, a couple of generations of LAN e-mail software has developed. Unfortunately, while these packages provide far more features than SMTP, they're often unable to communicate with other LAN e-mail packages or with mail services running on larger networks. Because LANs often originated at the departmental level without MIS directives on standardized software, this situation isn't uncommon.

Now that enterprise networking is becoming a corporate goal, network managers and systems integrators often need to integrate several different company LAN e-mail packages with each other as well as with mail services running on larger computer networks, particularly such corporate favorites as DEC's All-in-1 and IBM's PROFS and SNADS. In this section, you'll see how evolving industry standards could help resolve some of these present incompatibilities.

Simple mail transfer protocol (SMTP)

First there was simple mail transfer protocol (SMTP). SMTP was developed for the Defense Department's network as a "no frills" electronic messaging system that could transfer mail between two machines running on different networks, often connected via Telnet. Most Unix systems support SMTP although it's usually not used unless the networks are running TCP/IP.

SMTP consists of only 14 different commands described in a simple document (RPC 821). It's efficient but limited. While e-mail functions have expanded dramatically over the past few years, particularly in the very competitive LAN e-mail market, SMTP has remained frozen. There has also been a growing movement toward an international e-mail standard, one that will serve as a common language for proprietary mail systems that need to communicate.

The X.400 standard

Incompatible e-mail systems are a major headache for any network manager charged with the responsibility of creating an enterprise network. The CCITT has established a number of standards for electronic mail that have become part of the OSI model. The original CCITT X.400 recommendations were adopted in 1984 and a set of updates, now undergoing testing, were published in 1988. These 1988 changes (known in the industry as 1988 X.400) include the ability to mix text and nontext items, such as spreadsheet and graphics files, audio, and video information. Each of these items is tagged by a unique object ID. Vendors must obtain object IDs for their own multimedia data formats from ANSI and then publish this information. It will be a while before these new X.400 features start to appear in e-mail products.

Figure 12.2 illustrates an X.400-compliant message handling system (MHS). A network user interface to the mail system is known as a user agent (UA). This interface enables a network user to retrieve or send mail. The user agent constructs the X.400 "envelope" and places addressing information in the appropriate header fields. It looks up any required X.400 addresses and constructs distribution lists if they're requested. The user agent transmits its envelope containing headers and message to a message transfer agent (MTA).

Think of MTAs as post offices that offer typical postal services for electronic messages. They include the store-and-forward service of storing a message until a specified time before forwarding it to its destination. MTAs also perform error checking and ensure that the envelope and headers have been formatted correctly. The message transfer system (MTS) provides enhanced delivery services, such as verified delivery or nondelivery.

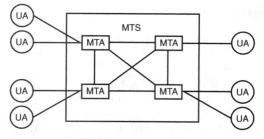

Figure 12.2 An X.400 message handling system.

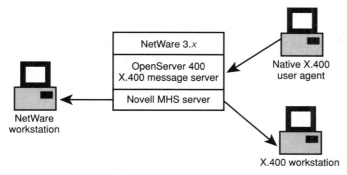

Figure 12.3 The Retix OpenServer 400 LAN X.400 MHS gateway.

The X.400 set of specifications include a P1 protocol, which describes the envelope. The X.400 envelope includes a header that contains such key information as the sender, recipient, subject references, and copy lists. User agents communicate with message transfer agents using the P2 protocol, which specifies how the messages are structured as well as delivery instructions.

A LAN e-mail user who is X.400-compliant or contains an X.400 gateway writes a message and attaches any additional required documents (spreadsheets, text files, etc.). The user indicates the recipient's X.400 address or the user agent looks it up. The message is transmitted via a user agent from MTA to MTA until it reaches the recipient's MTA.

IBM offers its X.400 open system message exchange (OSME) as part of its OSI communications subsystem. This software enables IBM mainframes running under MVS or VM to act as X.400 MTAs. Since IBM also offers its X.400 PROFS and DIOSS connection packages, which convert between X.400 format and PROFS and DIOSS formats, mainframe users can transmit an electronic message created under PROFS or DIOSS to an X.400-compliant e-mail system.

The LAN X.400 gateway server

Several companies, including Novell and Retix, have formed the X.400 application program interface association (APIA) in order to develop specifications for X.400-compliant APIs. These APIs can be used to create gateways to handle the protocol conversion and routing necessary to connect proprietary e-mail systems with X.400 systems, as shown in Figure 12.3.

Let's look at a specific X.400 gateway, Retix's OpenServer 400, and see how it operates in a LAN environment. First, it's necessary to clear up some confusing terminology. A few years ago, Novell began including Action Technologies' message handling service (MHS) with NetWare, which was an e-mail interface quite distinct from X.400's MHS. Recently, Novell has been doing all the developmental work on this product, currently renamed *mail handling service*; it's still not X.400-compatible.

Retix has developed an OpenServer 400 MHS X.400 gateway that allows NetWare LANs to exchange messages with other MHS LANs, as well as networks using X.400 messaging systems. The initial product offering consists of a client/server model. Client PC workstations run their own copies of Retix's user agent software as an application under Microsoft Windows. A message is transmitted to a dedicated message server running the MTA software.

Managing enterprise mail

The Internet Engineering Task Force has developed a standard for managing e-mail routers via SNMP. The mail and directory management (MAD-MAN) management information base (MIB) could make it easier to manage complex enterprise e-mail networks. A MADMAN MIB can tell a network manager whether a message transfer agent (MTA) is working, as well as how many messages are queued up in the MTA. It can also identify the age of the oldest message. This type of information can be used to trigger appropriate alarms. Proposed extensions to this standard would permit tracking of specific messages and displaying the average time a message is in a queue.

Linking e-mail gateways together

What about a situation where a company has substantial investments in not just IBM hosts running PROFS and DIOSS, but also DEC systems running All-in-1 and VMSmail? The company wants to link mail users on these systems with LAN and UNIX users. While the Retix/Novell MHS X.400 gateway can handle a small part of this problem, it can't solve the network manager's problem.

Touch Communications has developed a series of X.400 gateways known as WorldTalk/400. This product line enables e-mail users on AppleTalk, NetWare, and TCP/IP LANs to communicate with public messaging nets, such as AT&T Mail, MCI Mail, IBM's PROFS, and DEC's All-in-1. Touch uses a far different approach than Retix. It divides its messaging system into two key components: a gateway engine residing on a UNIX-based workstation and gateway modules residing on LAN file servers or mail servers. The LAN modules perform the addressing as well as the conversion from that LAN's e-mail format to the X.400 format. If several different LANs are linked to a single gateway engine running on a UNIX machine, then routers are used to link together the LANs. Figure 12.4 illustrates how Touch's system works.

The X.500 directory

One reason why Vines has been such a popular LAN with large organizations is its proprietary global naming service known as StreetTalk. Users on

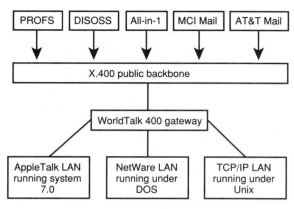

Figure 12.4   Touch Communications' WorldTalk gateway.

any subnetwork can communicate with each other over the Vines internet and use StreetTalk to locate any network addresses they need just as a phone user would turn to the white pages of a telephone book.

Vines' StreetTalk includes both white and yellow pages as "directory assistance." Users can search for recipients' nicknames, specific services, printers, and volumes regardless of where on the network they're located. In order for e-mail users on X.400 systems to communicate with each other transparently, there has to be a universal global directory performing the same function as StreetTalk.

If Carol Kennedy of ABC Corporation wants to communicate via an X.400 gateway with John Smith of XYZ Corporation, she needs to know Smith's network address on the XYZ LAN. When the CCITT's X.500 set of specifications for a global directory are finally implemented, Kennedy would simply look up Smith's address on her workstation screen. Similarly, a user on XYZ's PROFS system could look up the address of another company employee who happens to be on a network using DEC's All-in-1 software.

During 1989, the CCITT finalized its X.500 recommendation, scheduled to be written by 1992. One major issue still unresolved is how to restrict access control so a user from one network cannot access a directory for another network and have the ability to read, search, and modify anything on that network. One real worry is that junk mail companies will be able to compile detailed information on corporate phone numbers as well as individuals' names and addresses. We'll leave this problem to the experts developing the new set of recommendations.

There are many advantages to an enterprise-wide directory. For user services, it offers a single log-on to all authorized servers as well as a single log-on to all authorized services, such as high-speed printers and CD-ROMs on the network. Remote users enjoy the same access to services as local users.

For the network manager, an enterprise-wide directory offers improved network security, identification of services and their locations, and a single process to add users to LANs and e-mail services. It offers more efficient LAN administration, the ability to handle multiple e-mail services and synchronize their directories, and an easier way to balance loads between servers.

A user accesses an X.500 directory through a directory user agent (DUA). The DUA communicates using directory access protocol (DAP) with a directory system agent (DSA). The DSA decides whether it can handle this request or must broadcast it to other DSAs. The DSA can be distributed or centralized depending on how the directory database is created. DSAs use a directory system protocol (DSP) to broadcast messages to each other.

X.500 directories are designed in the form of a directory information tree (DIT). The root of this information tree is purely conceptual. Directly below it you might find countries, then organizations, and then virtually any designation that fits a particular company. Portions of the DIT are handled by individual DSAs. The directory is able to handle both white and yellow pages. The white pages provide name-to-address mappings, while the yellow pages enable a user to find another user or service by looking under specific business categories.

Figure 12.5 illustrates a typical DIT. Each entry has a set of attributes, which can be single or multivalued. A name entry might have the required attributes of first and last name, but other attributes, including the person's photograph, phone number, or social security number, are optional.

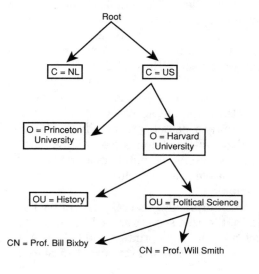

Figure 12.5 An X.500 directory information tree (*DIT*).

Banyan's enterprise network services (ENS)

Banyan has plans to become a major force in the enterprise directory market. Its StreetTalk is generally recognized as the most sophisticated and tested product on the market. The company realizes that only a small percentage of companies will become Vines users, yet many of these same companies need directory services. The result is a series of enterprise network services (ENS) products. ENS for NetWare, for example, enables a network manager with LANs running NetWare $2x$ or $3x$ to use StreetTalk directory services to manage these networks.

Banyan has plans to offer ENS products for most major platforms. The company was to become the "glue" that holds together enterprise networks containing LANs running a variety of network operating systems. It's ironic that ENS for NetWare can do something that even Novell can't do for its own customers—offer directory services for NetWare $2x$ and $3x$ LANs. Novell customers must upgrade to NetWare $4x$ to enjoy directory services.

Novell's name service

Novell has released its NetWare name service, which supports up to 400 servers and 5,000 users per domain with no limit to the number of domains that can be created. The utility program NETCON automatically copies users' names from the server in which they log on to all the other servers in that particular domain. If a server is out of service, it's resynchronized and has its names updated before it's allowed once again to become part of the network.

There are some severe limitations to Novell's first release. It lacks the hierarchical naming structure that characterizes StreetTalk. What this means is that user names can't identify the user, division, and department since they're limited to DOS's eight-character format. Novell's naming service currently doesn't include the "white pages" and "yellow pages" found under StreetTalk.

Novell's decision to offer directory services only to LANs running NetWare $4x$ leaves network managers who have enterprise networks running mixed NOS products with only two alternatives: upgrade all versions of NetWare to $4x$ or buy Banyan's ENS for NetWare. Many customers will continue to choose this alternative rather than be bullied into a NOS migration they don't want to make.

Network Security in an Enterprise Environment

There are two key aspects to network security on an enterprise network: protecting the network from unauthorized access and protecting the integrity of the data on that network. Ensuring both types of security requires a good deal of careful planning. Most studies have shown that approximately 80 percent of security violations come from employee error and equipment failure and the remaining 20 percent from unauthorized access.

An enterprise network offers an inviting target for unauthorized users because it usually encompasses several locations and involves remote communications; the size of an enterprise network often means that users are accustomed to seeing people they don't recognize walking around. Communications servers are often easy targets to penetrate, particularly if minimum precautions haven't been taken. Let's examine some specific actions that can be taken to increase security.

Password protection

At the most basic level, a network manager must require passwords and not allow everyone to log onto the network by simply typing USER or GUEST. Despite strenuous efforts by most network managers, many users rebel at the prospect of using nonsense passwords that are harder to guess and insist on using their names, the names of relatives, and even the names of their pets as passwords. Forcing users to change their passwords periodically will soon cause most users to run out of meaningful passwords that are easily guessed. Many users write out their passwords and leave them on top of their computers in case they forget them. Periodically walking the floor and removing these password notes will help alleviate this problem.

Still other users will place their passwords in batch files to automate the login process; unfortunately, this procedure makes it easy for snoopers to scan a DOS file and identify such passwords.

Oftentimes unauthorized users are legitimate network users who are curious about programs or files they have no right to examine. PCs in the personnel department, for example, might have access to a program that reveals employee salaries. Users should be encouraged to log off the network and never leave their PCs logged in when they leave their desks to go to meetings.

If there's real concern about stolen passwords, network managers can use dynamic passwords. There are new passwords for users as they log in. A hand-held remote password generator (RPG) device is used as well as special software running on the network. A user logging in is greeted with a challenge number. The user retypes the number along with a personal ID number into the RPG, which, in turn, generates a one-time-only password. The RPG is linked to the user's personal ID number and cannot be used by anyone else.

If there's a danger of unauthorized users tapping into data transmitted over a network, then data encryption might be the solution. Data that's encrypted is unreadable by anyone who does not have the required "key" to unlock the encryption.

There are two different types of encryption systems. *Public-key encryption* requires both creator and recipient of the data to have the same key to decrypt a file. Half of the key is made public and published in a directory while the other half remains private. A process known as *authentication* is required before encrypted files can be exchanged. After computer-to-computer communications that could involve matching 512 bytes of information, the authentication process is complete and the transmission and reception of encrypted files can begin.

Private key encryption uses a computer as a key distribution system in order to pass out keys to both sender and receiver. The sending computer sends a set of "credentials" consisting of user name, time date, etc., as a "ticket." This ticket is encrypted and then transmitted to the receiver computer, where the very same key is used to decrypt the message.

C2 security level

There has recently been a great deal of interest in the government's C2 level of security. This standard consists of six requirements for a trusted system:

- A clearly defined security policy must be enforced.
- All users and their access rights must be uniquely defined.
- All objects must be labeled with their security level.
- The system must track all security-relevant activities and secure this tracking information.
- The system must enforce security and be able to prove that enforcement.
- The system must be able to protect itself.

Controlling unauthorized access from remote users

One very easy way to control unauthorized access from remote network users is to use call-back modems. When a user logs in, the security device requests an access code, receives the code, and then hangs up. If the access code is correct, then the security device calls the user back at a preset telephone number.

Protecting Data Integrity on a Network

Most network managers have access to lists of intruder alerts but feel they're much too busy to spend time examining such lists, which can run dozens of pages long. Frequent illegal attempts to use a particular login could indicate a concerted effort to gain access to specific programs and files. One way to make it more difficult for unauthorized users is to restrict the login of users to specific times and days of the week. Certain employees obviously need access on weekends, but how many users need access after 11 P.M. or before 6 A.M.?

Terminated employees raise still another thorny security issue. How soon after an employee is terminated is the person's login eliminated from the network? A good policy is to make elimination of user's login a part of a checkoff list for the personnel department. A signature from MIS might be required before the termination process is complete.

Because data is such a valuable corporate asset, some companies restrict access not only at the directory level but also at the file level. Having diskless workstations in public areas also improves security because unauthorized users are unable to download information to disks even if they gain access.

Security dangers to data integrity

Data can be damaged a number of ways, both deliberately and accidentally. In this section you'll examine deliberate attempts to damage data, such as viruses and Trojan horses, and how to create a disaster recovery plan.

Viruses pose a major threat to network software, particularly enterprise networks that have extensive remote communications including bulletin boards. They're self-replicating bits of computer code that hide themselves within programs. They then attach themselves to other programs and ride them throughout the network. Once they get into a NetWare LAN's login directory, they can do immense damage by corrupting the login program itself.

There are many commercial programs available today to check for viruses. An enterprise network manager would be wise to not only install virus-checking software that checks directories before letting users log onto the network, but also initiate procedures concerning the uploading of software. Many network managers prefer to test all software on a small test network before adding it to the corporate network; this procedure minimizes the chances of a virus sneaking onto the network from commercial software (a horrible thought, but a reality in many documented cases).

Remember the story of the Trojan horse? This gift was taken into the gates of the city and resulted in that city's destruction when hidden soldiers exited the horse and opened the city gates to attack. Trojan horses are programs that appear to be innocent but contain hidden directions to do destructive things once they're activated. Often they're triggered by a specific date. One LAN manager turned on his network console one morning only to discover that overnight a Trojan horse program had activated itself (one minute after midnight) and the network was in severe trouble, with all kinds of corrupted files. Once again, a wise network manager exercises a good deal of caution before adding any software to the network, even when that software is free.

Disaster recovery plans

A major fire in a Los Angeles bank, the catastrophic destruction of a major telephone cable in Illinois . . . There are a lot of stories out there to convince most prudent network managers that enterprise network data is only as protected as the scope of the corporate disaster recovery plan. What happens if telephone lines go down for a company whose day-to-day operations require salespeople to dial in to access the customer database? What happens if a fire destroys all the data at a site? Are there duplicate files at another location? To be successful, disaster recovery plans must start very high up in an organization and have the full backing of senior management. Apple has a disaster recovery plan that includes bolted-down equipment and safe storage of data tapes to minimize damage. Some companies have contracted with satellite companies to arrange for backup VSAT arrangements should their main telephone lines go down. Most major disaster recovery plans have the following components:

Hot sites. Sites with computers and all necessary peripherals and software customers can use a number of services, such as IBM's business recovery services (BRS), to ensure continued operations even if the primary site is damaged.

Cold sites. These are sites with power supplies and all necessary hookups for equipment but no actual equipment. The customer supplies the equipment.

Telephone recovery services. Many companies use alternate long-distance carriers as part of their telephone recovery plan. These carriers store alternate routings for 800 numbers that can then be activated in the event of an emergency to the primary 800 numbers.

Off-site storage of backups. As part of a disaster recovery plan, companies must plan how to transport their tape backup to alternative sites. Some companies use the CTAM approach (Chevy truck access method), which consists of taking tape backups and transporting them via plane or truck to a second location for safekeeping. The electronic vaulting approach uses batch transmission of data via T1 or T3 leased lines to a remote location. Records are updated via a remote online log. If a disaster strikes, the network manager can determine the actual moment of failure by looking at this log. For companies who can't afford any delay in data recovery, a tool called a *shadow database* can be used. A company's entire database is automatically replicated and updated continually. If a disaster strikes at the corporate headquarters, the backup facility has a completely up-to-date system that can be used.

Preventing Disaster to a Wide Area Network

Enterprise networks have a wide area component. There are a number of things that can be done to prevent disaster across the WAN. Routers are by their very nature designed to redirect packets across alternate paths should one circuit fail. In such a situation, though, the heavy traffic across the alternative path will suffer congestion unless the network manager has planned enough bandwidth to handle the overload. It's also possible to use carrier services, such as frame relay, to deliver traffic over alternative routes.

What about providing an insurance policy for T-1 lines in case they fail? Some customers will invest in duplicate T-1 lines, but this is not only prohibitively expensive but also doesn't always ensure protection because alternative carriers often share the same cabling conduits. A natural disaster that destroys one T-1 line will often destroy a second T-1 line located adjacent to it. The alternative that often makes the most sense is to make arrangements with carriers that provide switched 56-Kbps, 64-Kbps, 384-Kbps, and even 1.5-Mbps transmission, such as AT&T's Accunet switched digital services. Customers pay only a per-minute fee when these circuits are used.

System fault tolerance

While mainframe and minicomputers have long offered elaborate protection against failure, until recently the *system fault tolerance* for LAN file servers—the degree to which a system is tolerant of faults without failing—has been limited. The system fault tolerance available for LAN file servers takes many different forms, including multiple file allocation tables, hot fixes, disk mirroring, and disk duplexing.

Multiple file-allocation tables

A file server's file allocation table (FAT) keeps track of where files are located. If the table becomes corrupted, the disk drive controller can't locate any files. One solution offered by some network operating systems, such as Novell's NetWare, is multiple file allocation tables. These copies are updated periodically. Should one version become corrupted, the file server can use another version.

The hot fix

Many file server disk controllers aren't intelligent enough to distinguish a bad sector from a good sector on a disk. The hot fix feature enables a disk drive controller to check an area before writing to it, and then dynamically remap that area if it contains bad sectors. A certain area of the file server's disk drive is designated as a redirection section. When a bad sector is discovered and mapped, the data that was to be written to that area is written to the redirection section instead.

Disk mirroring

What happens if a file server's entire hard disk fails? This would normally spell curtains for the network, but disk mirroring can keep the network up and running. An identical second hard disk drive is connected via cabling to the same hard disk controller that handles the original disk drive. This is a form of system fault tolerance known as *disk mirroring*. But what happens if the disk drive controller and not the disk drive goes bad? The LAN would still fail in such a situation.

Disk duplexing

Disk duplexing adds a second disk drive controller and cabling, as well as a second hard disk drive. This form of system fault tolerance provides greater protection than disk mirroring because it protects against failure of the hard disk drive controller.

Disk arrays

Many file server companies offer disk arrays as an alternative to disk duplexing. Redundant arrays of inexpensive drives (RAID) offers several levels of fault tolerance (levels 0 through 5 are defined, with several more under consideration). The higher the RAID level, generally the less disk space required for data protection. Multiple drives appear to a file server as a single very large logical drive. Data is driven to all drives simultaneously. A minimum of parity data collected under RAID is enough to reconstruct lost or corrupted data.

RAID 0. Known as *disk striping*, data is written across multiple disk drives. This RAID level offers optimized storage capacity and good performance at a low price. Because data is stored in blocks across all the disks in an array, this RAID implementation results in short read and write times for large files. The problem is that there's virtually no fault tolerance at this level. If one disk crashes, there is a hole in the data that cannot be filled.

RAID 1. This level offers the equivalent of disk mirroring and thus provides the most reliable configuration. Data is copied from one drive to the next drive. This approach provides data redundancy and fast read performance. The disadvantages to this approach are slower write performance and cost. Twice as much disk space is required than a company actually needs because of the disk mirroring.

RAID 2. This level spreads redundant data across multiple disks and includes bit and parity data checking. The disadvantage is high overhead (an entire drive of duplicate data).

RAID 3. This approach uses disk striping at a bit level. It requires a dedicated parity drive, but data can be accessed quickly. The disadvantage is that it can perform only one write operation at a time.

RAID 4. This approach consists of disk striping of data blocks, and requires a dedicated parity drive. Very fast read access is provided along with considerable fault tolerance. The disadvantage is slow write performance.

RAID 5. This performs disk striping of both data and parity information. This approach provides good fault tolerance, efficient performance, and is a very commonly used RAID level. The disadvantage of this level is slower write performance than found in RAID 0.

The most common form of RAID found today consists of disk mirroring (1), striping (3), and parity and striping (5). A Raid Advisory Board (RAB) is looking at new levels of RAID. One reason RAID is becoming so attractive for LAN managers is that its cost is dropping below $1 per megabyte. RAID 1 is included with NetWare and provides a base level of fault tolerance. Its disadvantage is a relatively small storage capacity. RAID 5 appeals to many network managers because it offers better cost benefits than RAID 0 and 1, and offers data redundancy across twice as many drives with more storage capacity.

What "rule of thumb" should enterprise network managers use when evaluating RAID products? For typical servers in the 2- to 5-gigabyte range, the cost benefits of RAID 5 are insignificant compared to the reliability offered by RAID 1. Also, a RAID 5 system can be 30% slower than a RAID 1 system. For larger storage needs, RAID 5 appears to be the RAID level of choice. It's much more efficient than RAID 3 in reading in transaction processing environments.

Some network managers are attracted to the cost benefits of a software implementation of RAID 5 that uses existing disks and hardware. One drawback to this approach is that the extra processing lowers performance of both the server and the disk array. A bus-based implementation of RAID 5 offers the fastest implementation of this level of protection. Figure 12.6 describes some differences among RAID implementations.

Summary

It's cost-effective to use a fax server, but there are some problems associated with it. It still isn't possible to deliver fax messages directly to an end user's workstation.

Linking together e-mail software running on different network platforms is becoming easier with the development of the CCITT X.400 family of e-mail protocol standards. Linking together users on networks running at different locations on different file servers requires a directory service. The CCITT has been developing an X.500 standard for directory services. IBM, AT&T, and DEC have all committed themselves to developing compatibility with both X.400 and X.500 standards.

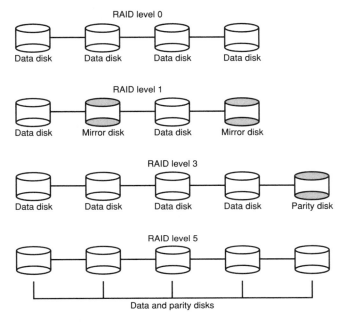

Figure 12.6 Some RAID implementations.

Security is crucial in an enterprise network, and it can take many different forms. You can prevent unauthorized users from accessing the network, and also protect the integrity of the data found on the network.

Disaster recovery planning is particularly important in an enterprise network environment. Wide area networks, in particular, present some significant security problems, such as finding alternatives for long-distance communications.

Seminars on LANs, Internetworking, and Enterprise Networks

Network Training

A character in Ernest Hemingway's *A Farewell to Arms* decides the world isn't a fair place at all and muses that the mysterious powers that be "throw you in the game. They don't tell you the rules. And when you break one, they kill you." Many network managers probably feel the same way. Having been responsible for developing coursework for and training network professionals, I can tell you that the quality of training in the industry varies widely. Here are some options you should consider:

Community colleges

Community colleges often offer evening training courses that are quite good and quite reasonable. One program in the Southern California area even offers certified NetWare engineer (CNE) training. The disadvantage of such programs is that they're time-consuming (once or twice a week for up to 16 weeks). Many network professionals can't budget the time.

Books and videos

Some people can learn from books, while others feel they need an instructor. There are some excellent training books available on CNE certification, LAN troubleshooting, the basics of administering NetWare, etc., but they do require a lot of self-discipline. Videos are advertised in leading LAN periodicals.

Sample books before purchasing them. Look through several books to see how different authors handle the same material. Do you like authors who give a lot of different examples, or do you prefer authors who take a single example and follow it through several different scenarios? What about reference material? Do some books have appendices that will prove valuable and justify their cost? How readable is the book?

Dealer and VAR training

If your company bought its networking equipment from a computer store or value added reseller (VAR), you might be in luck because both often offer introductory networking courses. There might be a minimum charge.

Seminars

Seminars vary in quality, but they offer an excellent way to pick up extensive knowledge in a limited amount of time. In this section I've listed several seminar companies that offer courses two, three, or four days in length. I've noted the courses that include "hands-on" opportunities.

This listing doesn't represent a personal endorsement. The best way to evaluate these courses is to review the outlines from brochures to look for depth rather than merely broad surveys that are too theoretical to be of practical value. Some companies will sell materials separately. While not ideal, it might be the solution if your training budget is really tight.

American Institute

437 Madison Avenue
23rd Floor
New York, NY 10022
800-345-8016

The American Institute offers seminars at six different technology centers (Chicago, Los Angeles, New York, San Jose, Toronto, and Herndon, Virginia). All courses are hands-on. The company has offered such relevant seminars as the following: data communications, troubleshooting Datacomm system networks, hands-on OSI (featuring GOSIP for those migrating to OSI), hands-on APPC, hands-on troubleshooting of local area networks, hands-on introduction to Novell NetWare, and hands-on internetworking (using bridges, routers, and gateways).

The American Research Group Inc.

P.O. Box 1039
Cary, NC 27512
919-380-0097

This company focuses on hands-on courses with some real depth. While prices change rapidly in this field, a typical five-day course on configuring Cisco router software runs $1,800. This particular class provides Cisco certification, an important consideration. Other courses in the LAN/WAN arena that the American Research Group have offered include those on the following topics: introduction to LANs, networking fundamentals, hands-on LAN wiring systems, hands-on fiber-optic cabling, integrating intelligent hubs, hands-on internetworking bridges and routers, ATM applications, introduction to network protocols, hands-on internetworking TCP/IP, SNA/LAN integration, hands-on SNMP, hands-on introduction to UNIX levels I and II, and hands-on UNIX internetworking.

Amdahl Corporation

1250 East Arques Avenue, MS 301
P.O. Box 3470
Sunnyvale, CA 94088-3470
800-233-9521

Amdahl offers several courses on LANs, WANs, and networks, as well as a whole set of courses on UNIX. There are also courses on NetView and data communications.

Business Communications Review

950 York Road, Suite 203
Hinsdale,IL 60521-2933
800-227-1234

BCR offers hands-on courses on local area networks and Ethernet LANs. Unusual courses include hands-on enterprise network troubleshooting, global networks, and fundamentals of wireless networking. The company has beefed up its internetworking area to include courses in internetworking with bridges and routers, LAN/WAN/LAN interconnection, hands-on wide area network design, hands-on TCP/IP internetworking, hands-on UNIX internetworking using TCP/IP, and simple network management protocol (SMNP). Several other UNIX courses are offered. BCR also offers telecommunications courses (its traditional area of expertise) on such topics as T-1, frame relay, ATM, and PBX technology.

Data-Tech Institute

Lakeview Avenue
P.O. Box 2429
Clifton, NJ 07015
201-478-5400

For several years this company has offered seminars on such topics as: lo-cal area networks, data communications, troubleshooting and maintaining a local area network, taking control of Novell NetWare for administrators and users, and troubleshooting Novell NetWare (fine-tuning for optimum network performance). The company also offers some telecommunications seminars.

Systems Technology Forum

10201 Lee Highway, Suite 150
Fairfax, VA 22030
800-336-7409

STF offers courses in such areas as network design, systems network archi-tecture (SNA), local area networks, and introduction to data communications.

Technology Exchange

Route 128
One Jacob Way
Reading, MA 01867
800-333-5177

Technology Exchange offers courses in such areas as UNIX, data com-munications and networks, local area networks, and computer network ar-chitecture. William Stallings has taught the LAN networks workshop.

Technology Transfer Institute

741 Tenth Street
Santa Monica, CA 90402
213-394-8305

TTI has always offered courses that are on the "cutting edge" of technol-ogy. They have offered courses in such areas as LANs, network management, network performance testing, ATM, TCP/IP, and OSI protocols. The company uses well-known industry figures such as James Martin and William Stallings.

Wave Technologies Training

111 Westport Plaza, Suite 300
St. Louis, MO 63146
800-828-2050

Wave Technologies has become the CNE course leader, with all kinds of special programs including what amounts to CNE summer camps. The company also offers telecommunications, network management, and internetworking courses, as well as videos.

Questions to Ask When Considering Seminar Training

- Can I have an outline of the course prior to enrolling?
- What is the practical experience of the instructor?
- Will hands-on training be provided?
- Can I bring a few of my actual real-life network problems with me to discuss with the instructor? Will there be time?
- What training materials will I take with me from this course?
- Can I have the telephone numbers of some people who have taken this specific course from this instructor?
- Is any free software offered with the class?
- Does your company offer a certification program?
- What happens if I'm not happy with the course?

B

Directory of Vendors

Bridges and Routers

3 Com Corporation
800-638-366

Advanced Computer Communications
800-444-7854

Allied Telesis
800-424-4284

Andrew Corporation
310-784-8000

Cisco Systems
800-553-6387

Combinet
800-967-6651

Digital Equipment Corporation (DEC)
800-343-4040

FiberCom
800-537-6801

Fibronics
617-826-0099

Hewlett-Packard
800-752-0900

IBM
800-426-2255

Kalpana
800-488-0775

Madge Networks
800-876-2343

Micom Communications Corporation
805-583-8600

Newbridge Networks
703-834-3600

Newport Systems
800-368-6533

Proteon Inc.
800-545-RING

Retix
800-255-2333

Rockwell Network Systems
800-262-8023

Shiva Corporation
800-458-3550

Telebit
800-TELEBIT

Timeplex Inc.
800-669-2298

Ungermann-Bass
408-496-0111

Wellfleet Communications
617-275-2400

Vitalink
800-767-4533

Xyplex
800-338-5316

Frame Relay Services

Ameritech
800-832-6328

MCI Communications
800-933-9029

AT&T
800-248-3632

NYNEX
914-644-5470

Bell Atlantic
800-422-0455

Pacific Bell
510-901-6498

BellSouth Telecommunications
404-529-6757

PacNet
206-232-9900

BT North America
800-872-7654

Southwestern Bell
800-992-2355

Cable & Wireless
703-790-5300

Sprint Corporation
800-877-4346

CompuServe
800-433-0389

US West Communications
800-328-2879

EMI Communications
800-456-2001

WilTel
800-364-5113

Internetworking

Learning never stops when it comes to internetworking. New topologies, new protocols, even new types of products (branch office routers, for example) are the rule and not the exception in this dynamic field. Books are usually good reference materials, but not good sources for late-breaking information. It's easy to forget that a book usually takes at least a year to write and then several months to publish. This appendix lists some key industry periodicals that are worth examining. Many are controlled publications that are offered free to people who are key decision makers, such as network managers.

Business Communications Review

BCR Enterprises
950 York Road
Hinsdale, IL 60521-2939

This monthly periodical offers excellent, in-depth articles on internetworking, with an emphasis on wide area network technology. ATM is covered very frequently, as are SMDS, frame relay, and routing technology. It's also a good place to keep up with ISDN and network management systems.

CIO

492 Old Connecticut Path
P.O. Box 9208
Framingham, MA 01701-9208

This monthly periodical is directed to chief information officers. It provides an enterprise-wide perspective on network problems and good information on the financial implications of network issues. Case studies are a regular part of this magazine.

Communications News

124 South First St.
Geneva, IL 60134

This monthly magazine provides short case studies of networks, with an emphasis on telecommunications. There are often special issues devoted to topics such as microwave and VSAT.

CommunicationsWeek

CMP Publications
600 Community Dr.
Manhasset, NY 11030

This is the single best source of information on the communications industry. In addition to covering the regional Bell operating companies (RBOCs) and network companies, this weekly offers special magazine inserts devoted to topics such as wide area networking. Excellent product guides with feature lists for such products as network management software are frequently offered.

ComputerWorld

375 Cochituate Rd.
Framingham, MA 01701-9171

This weekly publication provides excellent coverage of major network trends. It's unusual because it provides good coverage of mainframe and minicomputers in addition to LANs. It also has the largest number of want ads of any periodical listed in this section.

Data Communications

McGraw-Hill, Inc.
1221 Avenue of the Americas
New York, NY 10020

This monthly magazine is the best source of information on the technical aspects of data communications and integrated voice/data communications. Excellent articles are usually found on such difficult topics as ATM, frame relay, gateways, routers, and bridges. Want to know how something works? This is the place to look first.

InfoWorld

1060 Marsh Rd.
Suite C-200
Menlo Park, CA 94052

This weekly is particularly good for keeping up with the computer industry in general. It's also the place to look first for the hottest rumors.

LAN Magazine

12 West 21st St.
New York, NY 10010

This magazine is written at the elementary level, but it never talks down to its audience. It's an excellent place to begin to learn about internetworking topics such as frame relay, bridges, and routers.

LAN Times

McGraw-Hill, Inc.
7050 Union Park Center
Suite 240
Midvale, UT 84047

This biweekly publication is a must for anyone who wants to keep up with the internetworking field. It usually contains over 100 pages packed with information on the latest network products.

Network Computing

CMP Publications
600 Community Dr.
Manhasset, NY 11030

This monthly periodical provides excellent technical information on the current world of networking, with an emphasis on data networks.

Network World

375 Cochituate Rd.
Framingham, MA 01701-9171

This weekly publication is probably the single most important voice-oriented periodical in the internetworking area.

PC Week

P.O. Box 5826
Boston, MA 02199

This weekly publication provides more detailed information than *InfoWorld* on PC hardware and software. Its networking section always provides valuable information on new products.

Glossary

10BaseT   An IEEE standard for 10-Mbps transmission of an 802.3 CSMA/CD contention network using unshielded twisted-pair wire.

10Broad36   An IEEE standard for 10-Mbps transmission of an 802.3 CSMA/CD contention network using broadband coaxial cable.

10Base2   An IEEE standard for 10-Mbps transmission of an 802.3 CSMA/CD contention network using baseband thin coaxial cable. Also known as "cheapernet."

10Base5   An IEEE standard for 10-Mbps transmission of an 802.3 CSMA/CD contention network using baseband thick coaxial cable.

Accumaster   A UNIX-based workstation under AT&T's unified network management architecture that collects information from all major network components, including T-1 lines, LANs, PBXs, and mainframe computers.

active monitor   The workstation that assumes the role of monitor on a token-ring network.

active star   A topology in which a controller establishes active links to other star networks.

address bus   The bus carrying the address of the next instruction to be executed.

advanced program-to-program communication (APPC)   An IBM programming interface that enables programmers to write code that permits programs to communicate with each other.

all-routes broadcast frame   A frame transmitted to all network nodes by a workstation that's gathering routing information on a source routing network.

AppleTalk   Apple's suite of protocols for its local area network.

AppleTalk transaction protocol   A protocol that provides acknowledgment of a datagram's error-free delivery.

application process   A program residing within a specific application.

attached resource computer network (Arcnet)   A network scheme developed by Data Point and now licensed to several other vendors that features a 2.5-Mbps token bus LAN.

autoboot ROM   A ROM chip that enables diskless workstations to boot automatically from the file server.

automatic repeat request (ARQ)   Techniques used for error checking.

baseband coax   Coaxial cable that contains a single channel for carrying data at high speeds.

bit stuffing   A technique that ensures that no more than five consecutive 1-bits appear in a frame. 0-bits are inserted after five consecutive 1-bits to ensure that a flag field is unique. A receiving station reinserts the appropriate 1-bits.

broadband coax   Coaxial cable that carries multiple frequencies separated by guardbands.

broadband ISDN (BISDN)   ISDN running over broadband cabling can achieve speeds of over 600 Mbps. This new specification is still being developed.

bus   A data highway.

bus mastering   A technique in which the file server moves data from the network adapter cards to the disk without use of the host processor.

cache memory   Very-high-speed RAM chips used to keep a copy of the information last accessed from disk.

carrier sense multiple access with collision detection (CSMA/CD)   A media access method for a contention bus network.

cascaded bridges   Several bridges linked together in sequential order.

centralized   When control rests in one location.

central office (CO)   AT&T's term for its point of presence.

central processing unit (CPU)   That portion of a computer in which instructions are interpreted and executed.

cheapernet   A term used to describe 10Base2, an 802.3 version of Ethernet on thin coaxial cabling.

client-server   An arrangement in which one computer runs server software and acts as a server for a second networked computer that runs client software. The server "back end" finds the information requested and sends it, not the entire file, back to the client "front end" that contains a user-friendly user interface.

clones   A term used to describe computers that are compatible to well-known models, such as the IBM AT. These clones generally offer features not found on the computer they're emulating and usually cost less. Clone manufacturers use additional features and lower prices to break down customer resistance to buying a non-brand name. Compaq, a company known for its reliable products, is an exception because its IBM clones generally sell for a higher price than the products they emulate.

cluster controller   A mainframe device that serves as an interface between the host computer and its many peripherals. The cluster controller consolidates the many streams of data it receives from often slower transmitting devices.

common management information protocol (CMIP)   The network management protocol designed to run on an OSI network.

common management over TCP/IP (CMOT)   A protocol consisting of standard CMIP with a presentation layer that maps OSI layers 6 and below to TCP/IP.

connection-oriented service   Type 2 LANs use this service, which provides acknowledgment for error checking as well as flow control and error recovery.

connectionless service   Found on LANs because of their high transmission rate and high reliability. Type 1 LANs under this approach don't use the LLC layer for error checking, flow control, or error recovery. They rely on the transport layer protocols for these services.

Consultative Committee on International Telegraphy & Telephony (CCITT)   An international organization that develops telecommunications standards; the European influence is quite strong, and historically voice communications have dominated over data communications as the major focus.

contention   A network in which workstations must "contend" with each other to use the network.

control bus   The bus (highway) inside a computer that carries control information.

data bus   The bus (highway) inside a computer that carries data.

data circuit terminating equipment (DCE)   An interface for the DTE to connect it to a network. A modem is an example of a DCE.

data flow control layer   The SNA layer, corresponding roughly to the OSI model's session layer. It handles the establishment of half-duplex and full-duplex network sessions.

data stream protocol   An AppleTalk protocol that establishes the actual communications session.

data terminal equipment (DTE)   A network device such as a computer or terminal.

datagram delivery protocol   An AppleTalk protocol that addresses specific logical ports to different networks to establish the route a datagram will take.

demand protocol architecture (DPA)   A 3Com product that moves network transport protocols in and out of memory as needed.

digital access and cross-connect system (DACS)   A device designed to switch DS-0 channels from one T-1 span to another.

direct memory access (DMA)   A DMA controller determines the source and destination addresses of information to be retrieved and then orders the data bus to perform these read/write operations.

directory information tree (DIT)   The form taken by X.500 directories.

directory user agent (DUA)   The means by which a user accesses an X.500 global directory.

disk server   A computer with a hard disk that's partitioned into volumes for each user. Programs have file locking but not record locking in this networking scheme that predates file servers.

diskless workstation   A workstation that contains no disk drives. It uses an auto-boot ROM to boot from the network file server.

distributed processing Workstations do their own processing rather than have processing centralized in a mainframe or minicomputer.

distributed queue dual bus (DQDB) A protocol used by an IEEE 802.6 MAN; it describes a dual bus topology with traffic traveling in opposite directions.

draft international standard (DIS) A specification that has passed both the working paper and draft proposal steps toward becoming an international standard.

draft proposal (DP) A specification that has passed the working paper stage on its road toward becoming an international standard.

dual attached station (DAS) Nodes—usually more expensive multiuser and mainframe hosts—attached directly to an FDDI network.

dual protocol stack When computers contain two completely different protocol stacks so both protocols can be run.

dynamic router A router that uses sophisticated algorithms to route packets the optimal path.

echo protocol A protocol under AppleTalk that allows destination workstations to echo the contents of a datagram back to the source workstation.

element management system (EMS) Software designed to manage a specific network component under AT&T's unified network management architecture.

enterprise management architecture (EMA) DEC's network management architecture that's built on a proprietary platform with open interfaces to most major protocols.

Ethernet The first nonproprietary local area network hardware and software, developed during the late 1970s through a partnership of DEC, Intel, and Xerox. It featured a 10-Mbps CSMA/CD bus network with thick coaxial cabling.

exchange termination equipment (ET) Defined by ISDN as equipment that terminates the digital subscriber line or extended digital subscriber line in the local exchange. This usually consists of central office equipment.

extended industry standard architecture (EISA) A 32-bit bus structure retaining compatibility with circuit cards for the IBM PC and AT models and compatibles. This standard was developed by a group of IBM clone manufacturers as an alternative to the Micro Channel architecture.

extended super frame (ESF) A framing scheme advocated by AT&T that consists of 24 frames grouped together with an 8,000-bps FE channel to carry control information.

fast packet A T-1 enhancement that permits special T-1 multiplexers to allocate bandwidth dynamically.

fiber distributed data interface (FDDI) An ANSI standard for a 100-Mbps fiber-optic network.

file server A computer with a hard disk that's used to process network commands and store network files.

file transfer, access, and management (FTAM) A sophisticated file transfer program available under the application layer of the OSI model.

file transfer protocol (FTP)   A protocol for file transfer available with the transmission control protocol/internet protocol suite.

fileshare   Filesharing software available on a Macintosh running System 7.0 that enables network users to make all or part of their hard disk available to other users.

filter   A term used to describe the way a bridge examines a packet and decides, on the basis of its address tables, whether or not the packet should be forwarded.

focal point   Under IBM's NetView network management program, a focal point receives information from entry points located on various SNA network addressable units (NAUs).

forward   A term used to describe the action a bridge takes when it sends a packet on to the next bridge toward its ultimate destination.

fractional T-1 service (FT1)   Division of T-1 bandwidth into fractional units such as ⅛ (192 Kbps), ¼ (384 Kbps), or ½ (768 Kbps), which are sold to customers at a much more attractive rate than full T-1 service.

frame relay   A data link layer protocol that defines how variable-length data frames can be assembled using fast-packet technology.

front-end processor (FEP)   A software controllable controller that offloads overhead functions, such as polling devices, from the host so the host can concentrate on number crunching.

gateway   A device that connects networks with different network architectures. They use all seven layers of the OSI model and perform protocol conversion.

go-back-N continuous   An error checking method in which a station receives several frames before replying with a NAK or an ACK. If a station sent seven consecutive frames and an error was detected in frame four, the sending station would retransmit frames four through seven.

government open systems interconnect profile (GOSIP)   Set of protocols required so vendors will provide the government with compatible products.

heterogeneous LAN management (HLM)   A set of network management specifications developed jointly by IBM and 3Com to help network managers develop tools for monitoring, analyzing, and controlling the performance of heterogeneous LANs.

IEEE 802.3   The IEEE standard for a contention, CSMA/CD bus network.

IEEE 802.4   The IEEE standard for a noncontention, token-bus network that's physically a bus but logically a ring.

IEEE 802.5   The IEEE standard for a noncontention, token-ring network.

integrated services digital network (ISDN)   An evolving set of international standards for high-speed digital transmission of voice, data, and video over public phone lines at speeds up to 1.544 Mbps.

international communications architecture (IAC)   A feature of Macintosh's System 7.0 that permits applications to send data or commands to other applications located on either the same machine or a different machine on the network.

international standard (IS) A specification that has passed all steps and completed the process of becoming an international standard.

International Standards Organization (ISO) The organization responsible for adopting international communications standards; its open systems interconnect model is an example.

internet Two or more networks linked together with a router or gateway.

internet protocol exchange (IPX) The network protocol used on NetWare networks; it's derived from XNS but no longer compatible with this protocol.

jabber A condition in which a malfunctioning Ethernet node continuously transmits over the network.

jam signal A network signal generated to indicate that a collision has occurred on a local area network.

kernel The key administrative center for OS/2; it's responsible for scheduling, file management, memory management, and overall system coordination.

LAN Manager A network operating system designed by Microsoft to run under OS/2. Microsoft licenses this product to several companies, such as 3Com, AT&T, and Hewlett Packard, who develop their own versions.

LAN Manager/X A version of LAN Manager designed to work under UNIX.

line termination equipment (LT) Defined by ISDN as equipment located within the local exchange company's or common carrier's network in situations where lines must be extended beyond the normal range of a central office.

link test A 10BaseT function that constantly scans lines between nodes and hubs as well as lines between hubs for evidence of electrical activity to ensure the line is still functioning.

lobe The cable connecting a multistation access unit to a workstation.

local area network A group of microcomputers networked to share hardware and software resources at one physical location.

local area transport area (LATA) The geographic area governed by a local exchange company (LEC).

local exchange company (LEC) A local phone company, which might be an independent telephone company or one of the regional Bell companies.

LocalTalk Apple's cabling system for its AppleTalk local area network.

logical unit (LU) A logical port, a point of access that enables an SNA user to access the network.

management information base (MIB) A database containing key network management information for the SNMP protocol.

Manchester encoding A transmission scheme used by many networks. A negative voltage for the first half of but bit time represents the value 1, while a positive voltage followed by a transition to a negative voltage represents the value 0.

manufacturing automation protocol (MAP) A suite of protocols developed specifically for a manufacturing environment based on the OSI model, which incorporates the IEEE 802.4 standard for a token bus network. The major reason for MAP is to enable factory devices to communicate with each other and with network management tools. General Motors has been a leading advocate of MAP.

message control block (MCB)   The format used by a packet transmitted from the redirector to the NetBios interface.

message transfer agent (MTA)   An X.400 component that provides store-and-forward services as well as other "post office" functions.

message transfer system (MTS)   An X.400 component that provides enhanced delivery services, such as verified delivery or nondelivery.

metropolitan area network (MAN)   A network designed to link together diverse local networks into a city-wide net.

Micro Channel architecture (MCA)   An intelligent bus or data highway capable of handling high-speed traffic on an IBM PS/2 microcomputer.

MS-net   Microsoft's original network operating system that was licensed to vendors to modify and then repackage under their own names.

multistation access unit (MAU)   Wire centers with bypass circuitry found on IBM Token-Ring Networks.

name binding protocol   A protocol under AppleTalk that matches a workstation's server names with internet addresses.

named mailslots   A quick and dirty way for two processes to communicate without a full duplex, error-free channel. These mail slots can be accessed by name.

named pipe   Full-duplex traffic within a computer or between two computers under OS/2. Named pipes can be accessed like any sequential file.

NetBIOS   The network interface IBM provides with its PC LAN program. It has since became a de facto standard and is used by many network programs.

NetView   IBM's network management program that runs on an IBM mainframe.

NetWare   The leading network operating system. Novell has indicated that it wants this product to be hardware independent and serve as a platform for all leading network protocols.

network addressable unit (NAU)   An SNA component capable of being assigned an address and receiving or sending information to the host. NAUs include logical units (LUs), physical units (PUs), and system service control points (SSCPs).

network control program (NCP)   A PU running on a communications controller (FEP).

network driver interface specifications (NDIS)   Developed jointly by Microsoft and 3Com, this set of specifications provides an interface freeing higher-level protocols running under OS/2 from any concern of a network's hardware.

network file system (NFS)   A protocol developed by Sun Computer Corporation and used on a variety of UNIX-based systems.

network management protocol   A suite of network management protocols based on CMIP and used by AT&T in its unified network management architecture.

network management system (NMS)   A system that manages and controls a network and provides features such as network configuration, alerts, and traffic analysis.

network management vector transport (NMVT)   The protocol by which LANs can communicate with NetView.

nodes   Network workstations.

noncontention   A network where workstations don't have to contend for the right to use the network. A token bus and token ring are examples of a noncontention network.

open systems interconnect (OSI) model   A model for the interconnection of heterogeneous computer networks developed by the International Standards Organization.

Open Token Foundation (OTF)   An organization of vendors formed to promote interoperability in token-ring products.

operating system (OS)   A group of programs that manages a computer's hardware and software resources.

OS/2   An operating system designed for Intel 80286-, 80386-, and 80486-based IBM microcomputers and compatibles.

packet assembler/disassembler (PAD)   A device that provides protocol translation from a data stream's native protocol to X.25. At the destination end of the transmission, the X.25 packets are translated to whatever protocol is required.

passive star   A topology where cables branch out of a central block. There's no repeater functionality.

path control layer   The SNA layer that handles routing and flow control.

peer-to-peer relationship   Workstations on a network that share resources as equals.

physical control layer   The SNA layer corresponding roughly to the OSI model's physical layer. It provides specifications for serial connections between nodes and high-speed parallel connections between host computers and their front-end processors.

physical unit (PU)   A physical device or communication link and its associated software and microcode found on an SNA network. Examples of PUs are communication processors and cluster controllers.

pipe   A part of memory that serves a buffer with in and out buffers. Pipes serve as channels between two programs, and the information flows serially.

point of presence (POP)   Interface points within a LATA for an interLATA carrier such as AT&T, MCI, or Sprint.

premises distribution system (PDS)   AT&T cabling system.

presentation services layer   The SNA layer responsible for formatting, translation, and other services associated with the way data must look.

printer access protocol   An AppleTalk protocol that handles streaming tape sessions as well as streaming printer sessions.

process   A single program and all the computer resources required to run it, including memory areas, descriptor tables, and system support. An application is composed of many different processes.

protected mode   A mode under OS/2 in which key registers and memory are protected for each program running so a computer can be multitasking.

protocol   A set of rules or conventions followed by computers so they can communicate together.

queues   Memory areas serving as storage locations for information. Any process can open and write to a queue.

real mode   A mode under OS/2 that allows a program designed to run under MS-DOS to run. This mode limits the program to the hardware resources available under MS-DOS, including a maximum 640K of conventional RAM and 32 megabytes of hard disk storage.

redirector program   A program that enables programmers to write to it rather than to the NetBIOS. The redirector communicates with a file server using server message blocks (SMBs).

repository manager   The database that will hold network management information under IBM's SystemView.

ring   A network topology resembling a ring that's used for noncontention networks.

root bridge   Under the spanning tree algorithm, this is the bridge selected by its peers during the negotiation process as the one having the highest priority value and highest station address.

router   A protocol-specific device that connects two networks that might have completely different MAC layers.

routing table maintenance protocol (RTMP)   An AppleTalk protocol that keeps track of the number of bridges that must be crossed to send a datagram from one network to another.

runts   Packets that fall below the minimally acceptable packet size.

selective-repeat continuous   An error checking method in which a station saves all frames in sequence in a buffer so, if an error is detected in a particular frame (let's say frame 6), only that frame needs to be retransmitted.

semaphore   A kind of flag that exists in only two states—owned and not owned. Semaphores can signify that a process using a resource cannot be disturbed, synchronize two processes that need to communicate with each other, or signal when only one thread is in a position to monitor a particular event.

sequenced packet protocol (SPP)   An XNS protocol that ensures reliability above the simple datagram delivery of IDP by synchronizing sending and delivery with a full-duplex communication method. Packets are numbered, so lost or damaged packets can be requested and then retransmitted.

server message block (SMB)   The format used for a packet sent from the redirector to a file server.

service point   Under IBM's NetView, an interface to NetView from non-IBM equipment.

session   A link between two network addressable units (NAUs) under IBM's systems network architecture (SNA).

session protocol A protocol under AppleTalk that handles the correct sequencing of datagrams when they're out of order. It also takes responsibility for ensuring that datagrams are the correct size and that there are break points during conversations.

shared memory A way of handling memory so a portion of the host's memory is mapped to the NIC's memory.

signal quality error (SQE) A test on a 10BaseT network that ensures the cabling is functional and workstations are able to send and receive signals.

signaling System 7 (SS7) A series of CCITT recommendations that define the content and format of signaling messages necessary for transferring network control information.

simple mail-transfer protocol (SMTP) A mail service available under transmission control protocol/internet protocol.

simple network management protocol (SMNP) A network management protocol designed to run on TCP/IP networks.

single attached station (SAS) A network node attached to wiring concentrators on an FDDI network.

single-route broadcast frame When spanning tree topology is used with source routing bridges, this special frame is circulated once to ensure that only certain bridges in the network are configured to pass single-route frames.

solicit successor frame A frame sent on a token bus network that's designed to determine the next user of the network.

source explicit forwarding (SEF) A feature offered by many intelligent bridges that enables a network supervisor to assign internetwork access privileges by labeling specified addresses in a routing table as either accessible or inaccessible to specific users and groups.

source routing An algorithm used by IBM's Token Ring Networks where bridges keep track of routing tables and know the destination address of the workstation to which they want to route a packet.

source routing transparent bridge A bridge cable to forward both spanning tree and source routing frames.

spanning tree algorithm (STA) An approach to bridging multiple networks where more than one loop might exist. STA permits only one path to be active while alternate paths or loops are blocked.

star A network topology in which cables radiating out of a central computer or file server to each network workstation.

static router A router that requires the network manager to create routing tables. These tables remain static (unchanged) until the manager makes changes.

stop-and-wait ARQ An error checking method in which a computer transmits a frame of information and then waits to receive an acknowledgment (ACK) control code indicating that the frame arrived correctly.

StreetTalk A proprietary global naming service provided with the Vines network operating system.

subnetwork   Each network that's part of a larger internet.

super-frame   A standard DS-4 framing scheme consisting of 12 separate 193-bit frames. A framing bit is used to identify both the channel and the signaling frame.

synchronous optical network (SONET)   A set of specifications for synchronizing public communications networks and linking them together via high-speed fiber-optic links.

SystemView   IBM's network architecture plan for the 1990s that includes UNIX as well as SNA, TCP/IP, and OS/2.

system service control point (SSCP)   The network addressable unit (NAU) on an SNA network that provides the services necessary to manage a network or portion of a network. The SSCP resides in the virtual telecommunications access method (VTAM) control program on the host computer.

systems application architecture (SAA)   IBM's set of specifications for a common user interface and common programming interfaces for its entire line of computers.

systems network architecture (SNA)   IBM's network architecture, the layered suite of protocols found on its mainframes and minicomputers.

technical and office protocol (TOP)   A suite of protocols developed by Boeing and based on the OSI model. It was developed specifically to be used by the engineering, graphics, accounting, and marketing support functions in a manufacturing-oriented company such as Boeing.

Telnet   A virtual terminal service offered under transmission control protocol/internet protocol.

terminal adapter (TA)   An interface connecting non-ISDN-compatible equipment to an ISDN network.

terminal equipment-1 (TE-1)   Equipment that's ISDN-compatible and can be connected directly to an ISDN network.

terminal equipment-2 (TE-2)   Equipment that's not ISDN-compatible and requires a terminal adapter (TA) to attach to an ISDN network.

thread   An execution path within a process under OS/2. The three priority classes for threads under OS/2 are time-critical, regular, and idle time.

time-division multiplexing (TDM)   Multiplexing in which data streams are guaranteed a time slot whether they need it or not.

token   A specific bit pattern that indicates whether a token-ring or token-bus network is being used or not.

transaction services layer   The SNA layer corresponding roughly to the OSI model's application layer.

transmission control layer   The SNA layer that provides the pacing for data exchanges between NAUs. It also handles encryption when requested.

transmission control protocol/internet protocol (TCP/IP)   A suite of protocols designed for the Defense Department's network; it's now the basis of the nation's internet.

transparent bridging A technique used by all Ethernet and some token-ring networks where packets are forwarded "hop by hop" from bridge to bridge across the network.

two-way memory interleave A technique in which a computer accesses one bank of RAM chips while a second bank of RAM chips is being refreshed.

unified network management architecture (UNMP) AT&T's network management architecture, built on a foundation of OSI network management protocols.

UNIX A multiuser, multitasking operating system developed by Bell Labs; it's used on many large networks, particularly for scientific and university networks.

user agent Defined by X.400 recommendations as the component providing the X.400 envelope, all necessary headers, and addresses before forwarding a message to a message transfer agent.

virtual memory A computer's ability to use secondary storage to handle very large programs. The computer fools the programs into thinking there's more RAM than there really is by swapping different portions of the program from secondary storage to RAM as needed.

virtual telecommunications access method (VTAM) This program, running on an SNA network, coordinates communications between teleprocessing devices and application programs—managing the flow of data. It also permits the network's configuration to be changed while the network is running.

virtual terminal (VT) A virtual terminal protocol developed for the OSI model.

wait states A computer must pause if its RAM chips aren't fast enough, and "wait" until these chips are refreshed and ready to receive additional information.

wide area network (WAN) A network linking together networks located in other geographic areas.

working paper (WP) The very first stage a specification takes on the road toward becoming an international standard.

wrapping A procedure in which an FDDI network activates its second ring to bypass and isolate a failed node.

X.400 A series of recommendations drafted by the CCITT for protocols to provide international standards for electronic messaging.

X.500 A series of recommendations drafted by the CCITT for protocols to govern a universal directory.

Xerox network systems (XNS) One of the first network protocols and the basis of the original 3Com EtherShare and Novell NetWare network protocols.

zone Under AppleTalk's terminology, the logical grouping of networks on an internet.

zone information protocol An AppleTalk protocol that maps the network into a series of zone names.

Index

ABOUT THE AUTHOR

Stan Schatt is the author of *Linking LANs* and *Understanding Network Management* (both for McGraw-Hill) as well as over 20 other books, including the bestselling *Understanding Local Area Networks*, now in its 4th edition. As director of the LAN Service for Computer Intelligence InfoCorp., Schatt's primary responsibility is to track the LAN market and provide accurate analysis and forecasts for major LAN vendors. He is widely quoted in all the major trade publications, including *PC Week*, *Network World*, *Computerworld*, and *LAN Times*. Schatt holds a Ph.D. from USC and is former chairman of the Telecommunications Management Department of DeVry Institute of Technology.